Information Theory and Rate Distortion Theory for Communications and Compression

Synthesis Lectures on Communications

Editor
William Tranter, *Virginia Tech*

Information Theory and Rate Distortion Theory for Communications and Compression

Jerry Gibson

ISBN: 978-3-031-00552-7   paperback
ISBN: 978-3-031-01680-6   ebook

DOI 10.1007/978-3-031-01680-6

A Publication in the Springer series
SYNTHESIS LECTURES ON COMMUNICATIONS

Lecture #9
Series Editor: William Tranter, *Virginia Tech*
Series ISSN
Synthesis Lectures on Communications
Print 1932-1244   Electronic 1932-1708

Information Theory and Rate Distortion Theory for Communications and Compression

Jerry Gibson
University of California, Santa Barbara

SYNTHESIS LECTURES ON COMMUNICATIONS #9

ABSTRACT

This book is very specifically targeted to problems in communications and compression by providing the fundamental principles and results in information theory and rate distortion theory for these applications and presenting methods that have proved and will prove useful in analyzing and designing real systems. The chapters contain treatments of entropy, mutual information, lossless source coding, channel capacity, and rate distortion theory; however, it is the selection, ordering, and presentation of the topics within these broad categories that is unique to this concise book.

While the coverage of some standard topics is shortened or eliminated, the standard, but important, topics of the chain rules for entropy and mutual information, relative entropy, the data processing inequality, and the Markov chain condition receive a full treatment. Similarly, lossless source coding techniques presented include the Lempel-Ziv-Welch coding method. The material on rate Distortion theory and exploring fundamental limits on lossy source coding covers the often-neglected Shannon lower bound and the Shannon backward channel condition, rate distortion theory for sources with memory, and the extremely practical topic of rate distortion functions for composite sources.

The target audience for the book consists of graduate students at the master's degree level and practicing engineers. It is hoped that practicing engineers can work through this book and comprehend the key results needed to understand the utility of information theory and rate distortion theory and then utilize the results presented to analyze and perhaps improve the communications and compression systems with which they are familiar.

KEYWORDS

information theory, rate distortion theory, fundamental limits on communications, fundamental limits on compression

To Tyler Dean Gibson.

Contents

Preface

The disciplines of Information Theory and Rate Distortion Theory were introduced by Shannon more than 65 years ago, and there is likely little argument about the impact of these ideas on digital communication systems and source compression. However, to a broad audience of practicing engineers and even many deeply involved in digital communications and multimedia compression, the details of the connections may be obscure. Indeed, there are those who would vigorously debate the practical utility of information theory and rate distortion theory when it comes to the communications and compression systems that have evolved over the years.

One reason for the obscure connections of information theory and rate distortion theory to practice comes from the content of most courses on information theory and rate distortion theory, which often require considerable preparation in mathematics and statistics before they can be undertaken. As a result, many outstanding engineers do not have the time nor motivation to delve into these topics. Further, most research papers in the field do not address the specific connections of their models and results with practical systems, other than perhaps a cursory sentence or two stating some broad practical problem at a high level.

In order to provide greater access to these fields over the years, some authors have written textbooks aimed at undergraduate engineering students, and some very nice books have been written indeed. In fact, perhaps the very best of the group of widely accessible treatments of information theory is the early, and very influential, text by Abramson. However, these texts have not produced much momentum for the spreading of this knowledge. One reason is that the texts themselves almost always point out that information theory and rate distortion theory establish existence but do not provide design methods. Another reason is authors often say that the complexity required to achieve the performance promised by information theory and rate distortion theory is prohibitive, which certainly discourages students and practitioners alike.

To exploit the true promise of information theory and rate distortion theory, it is necessary to understand the theory at a basic level, which sounds obvious, and it is also necessary to understand deeply the particular physical problem being addressed. After all, fundamental limits in information theory and rate distortion theory require good models of the channels and sources, and meaningful mathematical expressions for distortion measures. Even the most advanced books on information theory and rate distortion theory acknowledge these facts, and unfortunately, often point to these requirements as bottlenecks to applications, not a particularly motivating way to convince students to master the material!

This book is very specifically targeted to problems in communications and compression, to providing the fundamental principles and results in information theory and rate distortion theory for these applications, and to presenting methods that have proved and will prove useful to analyze and to design real systems. We hope to achieve these goals with a minimum of mathematical preparation, although considerable mathematical prowess is still needed, while still maintaining the inherent elegance and power of the theory and the results.

The target audience for the book consists of graduate students at the master's degree level and practicing engineers. As a result, the technical and mathematical sophistication required is likely beyond all but the most advanced undergraduate. This book is not intended to replace a careful theorem/proof development of information theory and rate distortion theory; however, it will provide a strong leg up on mastering the material in such a more advanced course, though.

It is hoped that practicing engineers can work through this book and comprehend the key results needed to understand the utility of information theory and rate distortion theory to practical problems in communications and compression, and further, to be able to utilize the results presented to analyze and perhaps improve the communications and compression systems with which they are familiar.

For Readers/Students: There are proofs in this book, and indeed, some proofs are necessary to establish key practical results, to allow the reader to obtain some basic proficiency, and to (hopefully) retain the elegance of the information theoretic approaches to practical problems. All results in this book have great practical relevance or they would not be included.

For Instructors: The book can be used for either a one-semester or a one-quarter course on information theory and rate distortion theory. For a semester, the proofs of the Kraft inequality and the Asymptotic Equipartition Property can be covered and the achievability proofs for the Channel Coding Theorem and the Rate Distortion Theorem can be expanded in full from the outlines presented. For one quarter, some proofs might be curtailed and the sections on the Gaussian autoregressive source and conditional rate distortion theory can be skipped.

Jerry D. Gibson
Santa Barbara, California
December 15, 2013

CHAPTER 1

Communications, Compression and Fundamental Limits

This book is about Information Theory and Rate Distortion Theory specifically pointed toward the problems of communications and compression. Information theory and rate distortion theory were introduced by Shannon in his landmark 1948 paper [1, 2], wherein he presents the fundamental results concerning lossless (noiseless) source coding, channel capacity, and rate distortion theory and demonstrates the critical roles played by entropy and mutual information in establishing these fundamental limits. Shannon revisited fundamental limits on lossy source coding in his 1959 paper [3], wherein he coined the term rate distortion function, proved coding theorems, calculated $R(D)$ for several examples, derived what we now call the Shannon lower bound on $R(D)$, and noted the duality between a rate distortion function and a capacity cost function.

These papers produced Shannon's three theorems that establish fundamental limits on lossless source coding, the maximum rate that can be communicated over a channel with arbitrarily small error probability, and the minimum rate required to reproduce a source at a destination with specified average distortion.

In this book, we are concerned with the rather general, and likely familiar, block diagram shown in Fig. 1.1. While we have shown the Modulator and the Demodulator in the figure in order to emphasize the communications system orientation of this book, in what follows, we group these blocks with the input and output of the channel, respectively, as is commonly done in information theory. As a result, the channel models that we consider represent the combination of the Modulator, Physical Channel, and Demodulator.

The block diagram with this generalized Channel is shown in Fig. 1.2, and this diagram represents the communications (and compression) systems of interest throughout this book. This simple figure is far more informative than one might first imagine. The presence of the Source, Channel, and User blocks are explicit indications that we must have models for these entities appropriate for our particular application. Further, the Encoder is separated into Source Encoder and Channel Encoder blocks, and the Decoder is separated into Channel Decoder and Source Decoder blocks. One might think that separating these components would penalize system performance. However, Shannon's results show that as long as each component operates optimally, this separation does not cause poorer overall system performance. This result is called the Information Transmission Theorem or Separation Theorem, and we discuss this theorem later in this chapter.

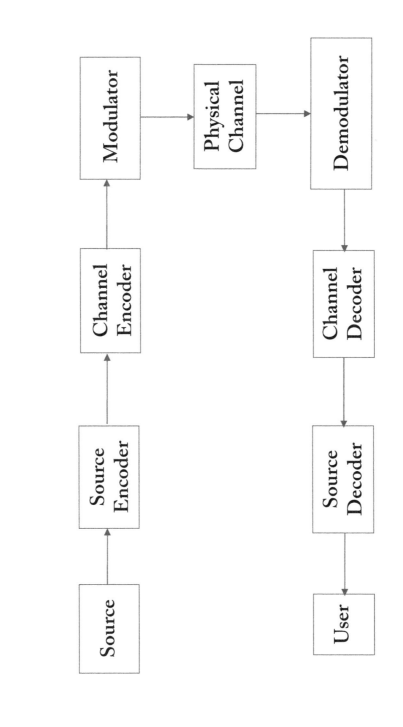

Figure 1.1: Communication system block diagram.

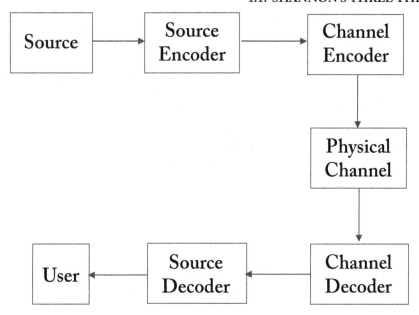

Figure 1.2: Simplified communication system block diagram.

Developing physically meaningful, but mathematically tractable models, for the Source, Channel, and User continues to be one of the great challenges to the impact of information theory and rate distortion theory on applications. While many significant results have been obtained, it is one goal of the current book to inform those most familiar with practical communication and compression systems of the basic underpinnings of information theory and rate distortion theory so they can contribute to this effort. The separation implied by the Information Transmission Theorem also has significant practical implications. Basically, if any component of the system is performing suboptimally, then the separate design of the source and channel encoder and decoders may not be the preferred approach. However, separation provides great flexibility and scalability to communication systems and is often the dominant design paradigm even when optimality is known not to hold.

In the next section, we introduce Shannon's three theorems and the underlying mathematical formulations for later exploration.

1.1   SHANNON'S THREE THEOREMS

What is normally called Shannon's first theorem is concerned with the minimum rate required to reproduce a discrete amplitude source exactly (losslessly) for storage or transmission. With this application in mind, we are primarily interested in the Source and Source Encoder/Decoder blocks in our general block diagram. Thus, we start with specifying a model for the Source. We model the discrete source as a random variable X with alphabet $\mathcal{X}$ with the letters of the alphabet $x \in \mathcal{X}$ having the probability $p(x) = P_r\{X = x\}$. Given this source model, the theorem is concerned with characterizing the minimum average codeword length of the codewords produced by a Source Encoder for lossless reconstruction at the Decoder. This minimum average codeword length corresponds to the minimum rate in bits per source letter possible for lossless transmission or storage of the given source (for the chosen source model).

Notationally, we investigate the properties of a lossless source code $C(x)$, which assigns a codeword of length $l(x)$ to each letter in the source alphabet. The average codeword length for this code is then $\bar{l}(C) = \sum_{x \in \mathcal{X}} p(x)l(x)$. Shannon's first theorem can be stated as [1, 4, 5]

Theorem 1.1 Shannon's First Theorem (Lossless Source Coding Theorem). *For a discrete source with entropy $H(X)$ bits/letter, there is a code C, such that for $\varepsilon \geq 0$, $\bar{l}(C) = H(X) + \varepsilon$. So, $\bar{l}(C) \geq H(X)$, where the entropy of the source X is $H(X) = -\sum_{x \in \mathcal{X}} p(x) \log p(x)$.*

This result says that the minimum rate or average codeword length required to represent a discrete source exactly for storage or transmission is lower bounded by the source entropy. Lossless source coding was originally called noiseless source coding (and still is many times) and some authors use the term data compaction [6].

Lossless source coding is used in many applications today, including still image compression, audio compression, and efficient data storage. Lossless source coding is the subject of Chapter 3 of this book. This chapter introduces the most common approaches to lossless source coding today, namely Huffman coding, Lempel-Ziv coding, and arithmetic coding, and provides approaches to analyzing lossless source coding performance, such as what is the penalty due to a mismatched source model.

Shannon's second theorem is concerned with the maximum rate that can be communicated over a channel with arbitrarily small error probability. This problem, if we follow the idea that source and channel coding can be separated, requires that we have a model for the Channel and then the theorem specifies the best that can be done by the Channel Encoder/Channel Decoder blocks.

A communications channel can be represented by $p(y|x)$, the transition probabilities between the input X and the output Y, and at the Channel Decoder, if $Y \neq X$, then an error occurs.

Shannon defined a quantity called the channel capacity as [1, 4, 5]

$$C = \max_{p(x)} I(X; Y) . \tag{1.1}$$

where $I(X;Y)$ is the mutual information between the two discrete random variables X and Y, representing the channel input and output respectively here, and given by

$$I(X,Y) = \sum_{x} \sum_{y} p(x,y) \log \frac{p(x,y)}{p(x)p(y)} \tag{1.2}$$

This quantity would merely be a mathematical expression were it not for Shannon's second theorem which gives it physical significance. Shannon's second theorem can be stated as [1, 4, 5]

Theorem 1.2 Shannon's Second Theorem (Channel Coding Theorem). *Capacity is the maximum rate that we can send information over the channel and recover the information at the output with arbitrarily small error probability.*

Notice that what this theorem says is that the distribution of the input to the channel can be chosen to maximize the rate that can be sent over the channel with asymptotically small error probability. In fact, the distribution of the input to the channel $p(x)$ is all that we have control over since the transition probabilities $p(y|x)$ are set by the model of the physical channel. Perhaps the most surprising implication of this theorem is that data can be transmitted over a noisy channel with an asymptotically small error probability. Such an achievement is possible, however, by letting the code block length get large, which is essentially trading delay for performance. This key idea is one of the many insights provided by Shannon.

Channel Capacity is defined in Chapter 4, wherein we also present properties of channel capacity, several example discrete channel models and the calculation of their capacity, several key techniques that can be used in evaluating capacity, a statement and outline of the proof of Shannon's second theorem for both discrete channels and Gaussian channels, and capacity results for several important Gaussian channels.

Turning our attention now to lossy source coding, we again separate source and channel coding operations, and so to explore fundamental limits on lossy source coding, we need a Source model and a model for the User. Generally, the source model is a distribution on the random variable that serves as input to the Source Encoder block. At the other end, the User model is some measure of fidelity between the input source and the output produced by the Source Decoder block, assuming by the Information Transmission Theorem that the Channel is ideal or that asymptotically small error probability can be achieved by the Channel Encoder/Decoder blocks for communication over the Channel.

For this introductory discussion, we let the input source be discrete and modeled by the probability mass function $p(x)$. We choose a distortion measure $\bar{d} = \sum_{x} \sum_{\hat{x}} p(x) p(\hat{x}|x) d(x,\hat{x})$ where $\hat{x}$ is the reproduction of the input letter x and $p(\hat{x}|x)$ indicates the probability of the reconstructed value $\hat{x}$ for the source input letter x. Notice that as a system designer, it is our job to select a meaningful source model and a useful distortion measure.

Shannon's third theorem allows us to characterize the minimum rate needed to represent the source with average distortion $\bar{d}$ less than or equal to some acceptable average distortion D. We can state the theorem as follows [3, 5, 7, 8]:

Theorem 1.3   Shannon's Third Theorem (Rate Distortion).   *The minimum rate that we can represent the source X with average distortion D is given by the rate distortion function, where $R(D)$ is defined as test channel $p(\hat{x}|x)$ such that*

$$R(D) = \min I(X; \hat{X}),$$
$$p(\hat{x}|x) : \bar{d} \leq D .$$

The key roles of the Source model $p(x)$ and the User model, through the chosen distortion measure, are clearly evident in this expression.

Chapter 5 is concerned with Rate Distortion theory and exploring fundamental limits on lossy source coding. Chapter 5 includes examples of the calculation of the rate distortion function for discrete sources and the important continuous Gaussian source, a discussion of the powerful and extremely useful Shannon lower bound, a development of reverse water-filling, a statement of Shannon's third theorem and an outline of its proof, and a development of rate distortion theory for sources with memory and composite sources.

It is important for the reader to note that it is the coding theorems that give the physical interpretation to the mathematical quantities of entropy and mutual information. This is one of the motivations for including some versions of the proofs in this book. A second reason for including proofs or outlines of the proofs is that the steps in the proofs provide insights into the power of information theory and Shannon's particular way of thinking about problems in communications and compression.

1.2   THE INFORMATION TRANSMISSION THEOREM OR SEPARATION THEOREM

As has been suggested earlier, the Information Transmission Theorem allows source coding and channel coding to be considered separately. In particular, the Information Transmission Theorem [6, 7] states that for a source with rate distortion function $R(D)$ and a channel of capacity C, then if $R(D) + \epsilon \leq C$ for some $\epsilon \geq 0$, the source can be reproduced with average distortion $D + \epsilon$. Further, if $C < R(D)$, the source cannot be reconstructed with average distortion $D + \epsilon$. For lossless source coding, the result is similar. For a discrete source with entropy H to be reconstructed losslessly at the output of a channel of capacity C, we must have $H < C$.

It is possible to prove this last converse result by using what is called the Data Processing Inequality. The Data Processing Inequality is enormously practical in its implications and so we introduce it here, although the careful development is deferred until later in the book. The basic idea is that for a communications system that behaves as a Markov chain, that is, $X \to Y \to Z$, from input to output, then $I(X;Y) \geq I(X;Z)$[5] . This mathematical result implies that the information between two quantities in the communication system, represented here by the mutual information, cannot increase as the signals propagate through the communication system. The Markov chain requirement is very reasonable for many if not most communication systems since it just says that there are no additional side channels or feedback channels. As we shall see later, the practical implications of the Data Processing Inequality are very interesting and perhaps surprising.

The Information Transmission Theorem states that source coding and channel coding can be accomplished separately without loss of performance if each step is performed optimally. Under this Separation Theorem, all of the quantities, namely, H, $R(D)$, and C, can be calculated separately, as indicated in Shannon's three theorems. Of course, it is well known that if there is a nonzero probability of error when using the channel even with error control coding, then the joint design of the source and channel coding schemes may prove advantageous. However, if we design a joint source/channel coding scheme, then we will likely lose performance if we switch either the source or channel coder in the joint design without reoptimization. Since it is often desirable, especially in today's multimedia-rich world, to send many different types of sources over a communications channel, joint source/channel coding can be considered to be a undesirable constraint. Such a limitation on joint source/channel coding is often called a lack of scalability for joint source/channel coding designs. In this book, we only examine the separation of source and channel coding as introduced by Shannon.

There is much to be learned from Shannon's original results and the separate source and channel coding implied by the Information Transmission theorem, and we will develop these fundamentals in the subsequent chapters of this book.

1.3   NOTES AND ADDITIONAL REFERENCES

The landmark papers by Shannon [1, 2, 3] defined the fields of Information Theory and Rate Distortion Theory, and Shannon's original intent was to provide theoretical underpinnings for communications and source compression problems. Although many papers and books and research studies have expanded the scope of information theory and rate distortion theory to other applications, it is the author's view that even with respect to the latest results in the field, the primary impact has been and continues to be on communications and compression. The reader should see the transcript of a recent panel with discussions concerning the impact and future directions of information theory and rate distortion theory [9].

Other than Shannon's papers, the citations in this chapter are to the books by Gallager [4], Cover and Thomas [5], Berger [7], and Blahut [6]. Berger's book is concerned with Rate Distortion Theory, while the others have broader coverage. There are many other excellent books on information theory and rate distortion theory, such as the early books by Abramson [10], Fano [11], and Ash [12] and the later books by Jelinek [13] and McEliece [14], all of which intend broader coverage of both fields. All of these books have their own approaches and unique perspectives and the author has found them to yield many insights. In addition to Berger, another important book that emphasizes rate distortion theory is the compact book by Gray [15].

Books with an information theoretic view of communications include Wozencraft and Jacobs [16] and Viterbi and Omura [17]. A very clear exposition of the algorithms used in data compression is contained in the book by Sayood [18].

Other references that have aided and influenced the author's views on the field include the compendia by Davisson and Gray [19], Slepian [20], and Verdu and McLaughlin [21]. The reader will also see the influence of Berger in this book from his excellent chapters on lossless source coding in Gibson, Berger, Lookabaugh, Lindbergh, and Baker [22] and his summary of the field of Rate Distortion Theory as of 1998 in the coauthored paper [8].

Numerous other references are cited throughout the book and all have contributed to the author's knowledge and enjoyment of the field.

CHAPTER 2

Entropy and Mutual Information

2.1 ENTROPY AND MUTUAL INFORMATION

As an initial source model, we consider a discrete random variable U that takes on the values $\{1, 2, \ldots, M\}$, where the set of possible values of U is often called the *alphabet*, denoted as $\mathcal{U}$, and the elements of the set are called *letters* of the alphabet [4, 5, 7]. Let $P_U(u)$ denote the probability assignment over the alphabet, then we can define the *self-information* of the event $U = u$ by

$$I_U(u) = \log \frac{1}{P_U(u)} = -\log P_U(u) \ . \tag{2.1}$$

The quantity $I_U(u)$ is a measure of the information contained in the event $U = u$. Note that the base of the logarithm in Eq. (2.1) is unspecified. It is common to use base e, in which case $I_U(\cdot)$ is in natural units (nats), or base 2, in which case $I_U(\cdot)$ is in binary units (bits). Either base is acceptable since the difference in the two bases is just a scaling operation. We will use base 2 in all of our work, and hence $I_U(\cdot)$ and related quantities will be in bits.

Example 2.1 A random variable U has a sample space consisting of the set of all possible binary sequences of length N, denoted $\{1, 2, \ldots, 2^N\}$. If each of these sequences is equally probable, so that $P[u] = 2^{-N}$ for all u, the self-information of any event $U = u$ is

$$\begin{aligned} I_U(u) &= -\log_2 P_U(u) = -\log_2\left(2^{-N}\right) \\ &= N \text{ bits} \ . \end{aligned}$$

The average or expected value of the self-information is called the *entropy*, also discrete entropy or absolute entropy, and is given by

$$H(U) = -\sum_{u=1}^{M} P_U(u) \log P_U(u) \ . \tag{2.2}$$

The following examples illustrate the calculation of entropy and how it is affected by probability assignments.

Example 2.2   Given a random variable U with the alphabet $\mathcal{U} = \{1, 2, 3, 4\}$ and probability assignments $P(1) = 0.8$, $P(2) = 0.1$, $P(3) = 0.05$, $P(4) = 0.05$, calculate the entropy of U. Compare the result to a random variable with equally likely values.

From Eq. (2.2),

$$H(U) = -\sum_{u=1}^{4} P_U(u) \log_2 P_U(u)$$
$$= -0.8 \log_2 0.8 - 0.1 \log_2 0.1 - 0.05 \log_2 0.05 - 0.05 \log_2 0.05$$
$$= 0.2575 + 0.3322 + .2161 + .2161$$
$$= 1.0219 \text{ bits}.$$

For equally likely values,

$$H(X) = -\log_2\left(\frac{1}{4}\right) = 2 \text{ bits}.$$

Lemma 2.3 $H_b(X) = (\log_b a) H_a(X)$.

Proof. Let $y = \log_a p$ so $a^y = p$. Then $y \log_b a = \log_b p$ or $\log_b p = \log_b a \log_a p$.   □

We now consider two jointly distributed discrete random variables W and X with the probability assignment $P_{WX}(w, x)$, $w = 1, 2, \ldots, M$, $x = 1, 2, \ldots, N$. We are particularly interested in the interpretation that w is an input letter to a noisy channel and x is the corresponding output. The *mutual information* over the joint ensemble is an important quantity defined by

$$I(W; X) = \sum_{w=1}^{M} \sum_{x=1}^{N} P_{WX}(w, x) I_{W;X}(w; x)$$
$$= \sum_{w=1}^{M} \sum_{x=1}^{N} P_{WX}(w, x) \log \frac{P_{W|X}(w|x)}{P_W(w)}. \tag{2.3}$$

By a straightforward manipulation of the mutual information, we get

$$I(W; X) = H(W) - H(W|X), \tag{2.4}$$

where $H(W|X)$ is the *conditional entropy*. Since entropy is a measure of uncertainty, we see from Eq. (2.4) that the mutual information can be interpreted as the average amount of uncertainty remaining about W after the observation of X.

Example 2.4   Here we wish to calculate the mutual information for the probability assignments (with $M = 2$ and $N = 2$)

$$P_W(1) = P_W(2) = \tfrac{1}{2} \tag{2.5}$$

and

$$P_{X|W}\left(1|1\right) = P_{X|W}\left(2|2\right) = 1 - p \tag{2.6}$$
$$P_{X|W}\left(1|2\right) = P_{X|W}\left(2|1\right) = p \ . \tag{2.7}$$

If we interpret W as the input to a channel X as the output, the transition probabilities in Eqs. (2.6) and (2.7) are representative of what is called a *binary symmetric channel* (BSC).

To calculate the mutual information, we use Eq. (2.3) so

$$
\begin{aligned}
I(W;X) = I(X;W) &= \sum_{w=1}^{2}\sum_{x=1}^{2} P_{WX}\left(w,x\right)\log\frac{P_{X|W}(x|w)}{P_X(x)}\\
&= \frac{1-p}{2}\log 2(1-p) + \frac{p}{2}\log 2p + \frac{p}{2}\log 2p + \frac{1-p}{2}\log 2(1-p)\\
&= 1 + (1-p)\log(1-p) + p\log p \ .
\end{aligned}
\tag{2.8}
$$

The mutual information given by Eq. (2.8) is plotted versus p in Fig. 2.1. The results of this example are discussed further in Sec. 4.1 in a communications context.

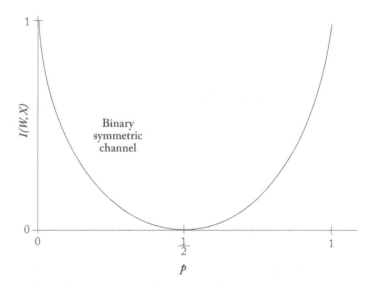

Figure 2.1: Mutual information for the BSC.

Example 2.5   Given the binary erasure channel (BEC) shown in Fig. 2.2, find an expression for the mutual information between the input and output $I(W;X)$ if $P_W(1) = P_W(2) = \frac{1}{2}$. The

BEC might be a good channel model for the physical situation where binary rectangular pulses are transmitted and the receiver makes a decision if the received pulse is much greater than or much less than the threshold, but asks for a retransmission of the received signal if the received signal is very near the threshold.

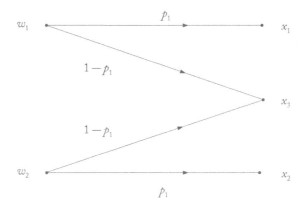

Figure 2.2: Binary Erasure Channel (BEC).

Using Eq. (2.3),

$$I(W; X) = \sum_{w=1}^{2} \sum_{x=1}^{3} P_{WX}(w, x) \log \frac{P_{W|X}(w|x)}{P_W(w)}$$
$$= \sum_{w=1}^{2} \sum_{x=1}^{3} P_{X|W}(x|w) P_W(w) \log \frac{P_{X|W}(x|w) P_W(w)}{P_X(x) P_W(w)} .$$

Now $P_W(1) = P_W(2) = 1/2$ and

$$P_X(k) = \sum_{w=1}^{2} P_{X|W}(k|w) P_W(w)$$
$$= \frac{1}{2}\{P_{X|W}(k|1) + P_{X|W}(k|2)\}$$

$$= \begin{cases} \frac{p_1}{2}, & k = 1 \\ \frac{2-2p_1}{2}, & k = 3 \\ \frac{p_1}{2}, & k = 2 . \end{cases}$$

Thus,

$$I(W;X) = \frac{p_1}{2}\log\left(\frac{p_1}{p_1/2}\right) + 0 + \left(\frac{1-p_1}{2}\right)\log\frac{1-p_1}{(2-2p_1)/2} + 0$$
$$+ \frac{p_1}{2}\log\frac{p_1}{(p_1/2)} + \left(\frac{1-p_1}{2}\right)\log\frac{1-p_1}{(2-2p_1)/2}$$

If $p_1 = p$,

$$I(W;X) = p + (1-p)\log\frac{2(1-p)}{2-2p} = p \text{ bits .}$$

Example 2.6 For the discrete memoryless channel (DMC) in Fig. 2.3 with $P_W(0) = \frac{1}{3}$ and $P_W(1) = \frac{2}{3}$, find $H(W)$ and $H(W|X)$. What is the mutual information for this channel and input probability assignment?

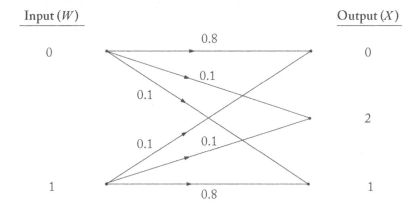

Figure 2.3: Discrete Memoryless Channel (DMC).

For $P_W(0) = 1/3$, $P_W(1) = 2/3$,

$$H(W) = -\frac{1}{3}\log\frac{1}{3} - \frac{2}{3}\log\frac{2}{3}$$
$$= 0.528 + 0.390 = 0.918 \text{ bits/letter .}$$

$$H(W|X) = -\sum_{x=1}^{3}\sum_{w=1}^{2} P_{W|X}(w|x)P_X(x)\log P_{W|X}(w|x) .$$

But

$$P_{W|X}(j|k) = \frac{P_{X|W}(k|j)P_W(j)}{P_X(k)} ,$$

so

$$H(W|X) = -\sum_{x=1}^{3} \sum_{w=1}^{2} P_{X|W}(x|w) P_W(w) \log P_{W|X}(w|x) .$$

Now from Fig. 2.3 we find that

$$P_X(0) = 0.8\left(\frac{1}{3}\right) + 0.1\left(\frac{2}{3}\right) = \frac{1}{3} ,$$

$$P_X(1) = 0.8\left(\frac{2}{3}\right) + 0.1\left(\frac{1}{3}\right) \cong 0.567 ,$$

$$P_X(2) = 0.1\left(\frac{1}{3}\right) + 0.1\left(\frac{2}{3}\right) = 0.1 .$$

Therefore,

$$\begin{aligned}
H(W|X) &= -(0.8)\left(\frac{1}{3}\right)\log 0.8 - (0.1)\left(\frac{1}{3}\right)\log\left(\frac{0.1}{1.7}\right) - (0.1)\left(\frac{1}{3}\right)\log\left(\frac{1}{3}\right) \\
&\quad - (0.1)\left(\frac{2}{3}\right)\log(0.2) - (0.8)\left(\frac{2}{3}\right)\log\left(\frac{1.6}{1.7}\right) - (0.1)\left(\frac{2}{3}\right)\log\left(\frac{2}{3}\right) \\
&= 0.086 + 0.136 + 0.0528 + 0.155 + 0.0465 + 0.039 \\
&= 0.5153 \text{ bits/letter} .
\end{aligned}$$

Therefore, from Eq. (2.4)

$$I(W; X) \cong 0.918 - 0.515 = 0.403 \text{ bits/letter} .$$

There are several key properties of entropy and mutual information. We state two of these properties here without proof.

Property 2.7   Let U be a random variable with possible values $\{u_1, u_2, \dots, u_M\}$. Then

$$H(U) \leq \log M \tag{2.9}$$

with equality if and only if the values of U are equally likely to occur.

Example 2.2 illustrates Property 2.7.

Property 2.8   Let W and X be jointly distributed random variables. The mutual information between W and X satisfies

$$I(W; X) \geq 0 \tag{2.10}$$

with equality if and only if W and X are statistically independent.

We now explore various expressions for mutual information which can be useful in communications and compression system analyses.

We expand the mutual information in two ways involving entropy and conditional entropy,

$$I(X, Y) = H(X) - H(X|Y) \tag{2.11}$$
$$= H(Y) - H(Y|X) . \tag{2.12}$$

where the joint entropy of a pair of discrete random variables $(X, Y) \sim p(x, y)$ is defined as

$$H(X, Y) = -\sum_{x \in X} \sum_{y \in Y} p(x, y) \log p(x, y) \tag{2.13}$$
$$= -E \log p(X, Y) . \tag{2.14}$$

and with the conditional entropy $H(Y|X)$ defined as

$$H(Y|X) = \sum_{x \in X} p(x) H(Y|X = x) \tag{2.15}$$
$$= -\sum_{x \in X} p(x) \sum_{y \in Y} p(y|x) \log p(y|x) \tag{2.16}$$
$$= -\sum_{x} \sum_{y} p(x, y) \log p(y|x) \tag{2.17}$$
$$= -E_{p(x,y)} \log p(Y|X) . \tag{2.18}$$

Expanding the joint entropy of (X, Y)

$$H(X, Y) = H(X) + H(Y|X) . \tag{2.19}$$

Proof.

$$H(X, Y) = -\sum_{x} \sum_{y} p(x, y) \log p(x, y)$$
$$= -\sum_{x} \sum_{y} p(x, y) \left[\log p(y|x) + \log p(x) \right]$$
$$= H(X) + H(Y|X) .$$

using the definitions. We can also show that $H(X, Y) = H(Y) + H(X|Y)$.

The definition of joint entropy and conditional entropy are natural since the joint entropy equals the entropy of one plus the conditional entropy of the other. We can continue the development as above to get the conditional version of the joint entropy

$$H(X, Y|Z) = H(X|Z) + H(Y|X, Z) . \tag{2.20}$$

Note that since $H(X, Y) = H(X) + H(Y|X) = H(Y) + H(X|Y)$, we have $H(X) - H(X|Y) = H(Y) - H(Y|X)$. $\qquad \square$

This equality is useful in evaluating mutual information for various applications.

A property that is used often in communications system analysis is that conditioning cannot increase entropy (this is often stated as conditioning decreases entropy). This result follows from the expression for mutual information since $I(X;Y) = H(X) - H(X|Y) \geq 0$, so

$$H(X|Y) \leq H(X) \tag{2.21}$$

with equality if and only if X and Y are statistically independent.

The averaging is important for this property since individual terms may be greater, that is, $H(X|Y = y)$ may be greater than $H(X)$.

2.2 CHAIN RULES FOR ENTROPY AND MUTUAL INFORMATION

We can build on the prior expressions and consider a set of jointly distributed random variables $X_1, X_2, \ldots, X_n$ drawn according to $p(x_1, x_2, \ldots, x_n)$ [4, 5]. Their joint entropy is the sum of the conditional entropies as follows

$$H(X_1, X_2, \ldots, X_n) = \sum_{i=1}^{n} H(X_i | X_{i-1}, \ldots, X_1) . \tag{2.22}$$

It is straightforward to justify this result since we already know that

$$H(X_1, X_2) = H(X_1) + H(X_2|X_1) ,$$

so,

$$
\begin{aligned}
H(X_1, X_2, X_3) &= H(X_1, X_2) + H(X_3|X_1, X_2) \\
&= H(X_1) + H(X_2|X_1) + H(X_3|X_1, X_2) \\
\vdots \quad &= \quad \vdots \\
H(X_1, X_2, \ldots, X_n) &= H(X_1) + H(X_2|X_1) + \ldots + H(X_n|X_{n-1}, \ldots, X_1) \\
&= \sum_{i=1}^{n} H(X_i|X_{i-1}, \ldots, X_1)
\end{aligned}
$$

where for $i = 1$, we define $H(X_1|X_0) = H(X)$.

The chain rule can also be obtained by starting with the basic definition of joint entropy and using conditional probabilities to expand the joint probability mass function. This is left as an exercise.

An important bound that finds wide applicability in practical applications is what is called the independence bound on entropy. For $X_1, X_2, \ldots, X_n$ with probability mass function

$p(x_1, x_2, \ldots, x_n)$, we have by the chain rule for entropies,

$$H(X_1, X_2, \ldots, X_n) = \sum_{i=1}^{n} H(X_i | X_{i-1}, \ldots, X_1)$$

$$\leq \sum_{i=1}^{n} H(X_i)$$

since conditioning cannot increase entropy.

X and Y given Z is defined by

An important result that follows from the chain rule for entropy is that the mutual information also satisfies a chain rule. To see this we write

$$
\begin{aligned}
I(X_1, \ldots, X_n, Y) &= H(X_1, \ldots, X_n) - H(X_1, \ldots, X_n | Y) \\
&= \sum_{i=1}^{n} H(X_i | X_{i-1}, \ldots, X_1) - \sum_{i=1}^{n} H(X_i | X_{i-1}, \ldots, X_1, Y) \\
&= \sum_{i=1}^{n} [H(X_i | X_{i-1}, \ldots, X_1) - H(X_i | X_{i-1}, \ldots, X_1, Y)] \\
&= \sum_{i=1}^{n} I(X_i, Y | X_{i-1}, \ldots, X_1)
\end{aligned}
$$

and thus we have

$$I(X_1, X_2, \ldots, X_n, Y) = \sum_{i=1}^{n} I(X_i, Y | X_{i-1}, \ldots, X_1) . \tag{2.23}$$

We will see later that this can be used with an independence assumption or a Markov chain property to simplify further.

2.3 DIFFERENTIAL ENTROPY AND MUTUAL INFORMATION FOR CONTINUOUS RANDOM VARIABLES

Thus, far we have defined the entropy and mutual information only for discrete random variables [4, 5, 7]. Since continuous random variables play an important role in many applications, particularly as source models for lossy source compression, we develop these definitions and properties here.

Given an absolutely continuous random variable U with probability density function (pdf) $f_U(u)$ we define the *differential entropy* of U as

$$h(U) = -\int_{-\infty}^{\infty} f_U(u) \log f_U(u) \, du . \tag{2.24}$$

Although this expression appears quite similar to the expression for the entropy of a discrete random variable, there is a significant difference between the interpretations of absolute or discrete entropy and differential entropy. While $H(U)$ is an absolute indicator of "randomness," $h(U)$ is only an indicator of randomness with respect to a coordinate system: hence the names "absolute entropy" for $H(U)$ and "differential entropy" for $h(U)$. The following example illustrates the calculation of differential entropy and its property of indicating randomness with respect to a coordinate system.

Example 2.9   Consider an absolutely continuous random variable with uniform pdf

$$f_U(u) = \begin{cases} \dfrac{1}{a}, & \dfrac{-a}{2} \le u \le \dfrac{a}{2} \\ 0, & \text{elsewhere .} \end{cases} \tag{2.25}$$

(1) Let $a = 1$ in Eq. (2.25) and find $h(U)$. Then

$$h(U) = -\int_{-1/2}^{1/2} \log 1 \, du = 0 . \tag{2.26}$$

(2) Let $a = 32$ and find $h(U)$. We have

$$h(U) = -\int_{-16}^{16} \tfrac{1}{32} \log \left(\tfrac{1}{32} \right) du = 5 . \tag{2.27}$$

(3) Finally, let $a = \tfrac{1}{32}$ and find $h(U)$. Here

$$h(U) = -\int_{-1/64}^{1/64} 32 \log(32) \, du = -5 . \tag{2.28}$$

The fact that differential entropy is a relative indicator of randomness is evident from these three special cases of the uniform distribution. Clearly, $h(U)$ is not an absolute indicator of randomness, since in case (3) $h(U)$ is negative, and negative randomness is difficult to interpret physically! The "reference"distribution is the uniform distribution over a unit interval, with "broader" distributions having a positive entropy and "narrower" distributions having a negative differential entropy.

Example 2.10   Calculate the differential entropy for an absolutely continuous random variable with pdf $f_U(u) = (1/\alpha)e^{-u/\alpha}, 0 < u < \infty$, and $f_U(u) = 0$ for $u \le 0$.
Given the pdf

$$f_U(u) = \begin{cases} \tfrac{1}{\alpha} e^{-u/\alpha}, & u > 0 \\ 0, & u \le 0, \end{cases}$$

from Eq. (2.24) with $\log = \ln$,

$$
\begin{aligned}
h(U) &= -\int_0^\infty \frac{1}{\alpha} e^{-u/\alpha} \log\left[\frac{1}{\alpha} e^{-u/\alpha}\right] du \\
&= -\int_0^\infty \frac{1}{\alpha} e^{-u/\alpha}\left\{-\log\alpha - \frac{u}{\alpha}\right\} du \\
&= \log\alpha + \left(\frac{1}{\alpha^2}\right)\alpha^2 e^{-u/\alpha}\left[-\frac{u}{\alpha} - 1\right]\Big|_0^\infty \\
&= \log\alpha + 1 = \log e\alpha .
\end{aligned}
$$

The Laplacian density is often used as a model for transform coefficients for head and shoulders images and video in transform-based compression techniques that dominate still image and video coding today. In the next example, we find the differential entropy for a Laplacian random variable.

Example 2.11 Show that the differential entropy for a random variable U with the Laplacian pdf $f_U(u) = (1/\sqrt{2})e^{-\sqrt{2}|u|}$ for $-\infty < u < \infty$ is given by $h(U) = \log(e\sqrt{2})$. By direct substitution into Eq. (2.24), (using $\log_2$)

$$
\begin{aligned}
h(U) &= -\int_{-\infty}^\infty \frac{1}{\sqrt{2}} e^{-\sqrt{2}|u|} \log \frac{1}{\sqrt{2}} e^{\sqrt{2}|u|} du \\
&= \frac{-1}{\sqrt{2}}\int_{-\infty}^\infty \left\{ e^{-\sqrt{2}|x|}\log\left(\frac{1}{\sqrt{2}}\right) + e^{-\sqrt{2}|u|}(-\sqrt{2}|u|)\log_2 e\right\} du \\
&= -\log\frac{1}{\sqrt{2}} + (\log_2 e)2\int_0^\infty xe^{-\sqrt{2}x} dx \\
&= -\log\frac{1}{\sqrt{2}} + \log_2 e = \log_2 e\sqrt{2} .
\end{aligned}
$$

It is often assumed that sources have a Gaussian distribution and so we next turn our attention to finding the differential entropy of a Gaussian distribution.

Example 2.12 Given a Gaussian source U with mean μ_s and variance σ_s^2, find an expression for its differential entropy $h(U)$. Use Eq. (2.24) and base 2 logs,

$$
\begin{aligned}
h(U) &= -\int_{-\infty}^\infty f_U(u)\log f_U(u) du \\
&= -\int_{-\infty}^\infty \frac{1}{\sqrt{2\pi}\sigma_s} e^{-(u-\mu_s)^2/2\sigma_s^2} \log \frac{1}{\sqrt{2\pi}\sigma_s} e^{-(u-\mu_s)^2/2\sigma_s^2} du \\
&= -\int_{-\infty}^\infty \frac{1}{\sqrt{2\pi}\sigma_s} e^{-(u-\mu_s)^2/2\sigma_s^2}\left\{\log\frac{1}{\sqrt{2\pi}\sigma_s} - \frac{(u-\mu_s)^2}{2\sigma_s^2}\log_2 e\right\} du
\end{aligned}
$$

$$= \log \sigma_s \sqrt{2\pi} + \frac{\log_2 e}{2\sigma_s^2} \frac{1}{\sqrt{2\pi}\sigma_s} \int_{-\infty}^{\infty} (u - \mu_s)^2 e^{-(u-\mu_s)^2/2\sigma_s^2}\, du$$

$$= \log \sigma_s \sqrt{2\pi} + \frac{\log_2 e}{2\sigma_s^2} \sigma_s^2$$

$$= \log \sigma_s \sqrt{2\pi} + \frac{1}{2} \log_2 e$$

$$= \log \sigma_s \sqrt{2\pi e} .$$

The importance of the Gaussian distribution for analyzing source coding problems and finding appropriate performance bounds is emphasized by the following result.

Theorem 2.13 ([1], [4])..   *For any absolutely continuous random variable ξ, the pdf that maximizes the differential entropy*

$$h(\xi) = -\int f_\xi(\xi) \log f_\xi(\xi)\, d\xi \tag{2.29}$$

subject to the constraint that

$$\int_{-\infty}^{\infty} \xi^2 f_\xi(\xi)\, d\xi \le \sigma_{\max}^2 \tag{2.30}$$

is

$$f_\xi(\xi) = \frac{1}{\sqrt{2\pi}\sigma_{\max}} e^{-\xi^2/2\sigma_{\max}^2} . \tag{2.31}$$

Proof. This result can be proved in several ways, including calculus of variations [1], relative entropy (introduced later), and Jensen's inequality [14]; however, an alternative method is used here [4]. Let $f_\eta(\eta)$ be an arbitrary pdf that satisfies the constraint in Eq. (2.30), and let $f_\xi(\xi)$ be given by Eq. (2.31). Then

$$-\int_{-\infty}^{\infty} f_\eta(\alpha) \log f_\xi(\alpha)\, d\alpha = \int_{-\infty}^{\infty} f_\eta(\alpha) \left\{ \log \sqrt{2\pi}\sigma_{\max} + \frac{\alpha^2}{2\sigma_{\max}^2} \log e \right\} d\alpha$$

$$= \frac{1}{2} \log 2\pi e \sigma_{\max}^2 . \tag{2.32}$$

Now consider

$$h(\eta) - \frac{1}{2} \log 2\pi e \sigma_{\max}^2 = \int_{-\infty}^{\infty} f_\eta(\alpha) \log \frac{f_\xi(\alpha)}{f_\eta(\alpha)}\, d\alpha$$

$$\le \log e \int_{-\infty}^{\infty} f_\eta(\alpha) \left[\frac{f_\xi(\alpha)}{f_\eta(\alpha)} - 1 \right] d\alpha = 0 , \tag{2.33}$$

where the inequality follows from the fact that $\log \beta \leq (\beta - 1) \log e$. Thus,

$$h(\eta) \leq \tfrac{1}{2} \log 2\pi e \sigma_{\max}^2 \qquad (2.34)$$

with equality if and only if $f_\xi(\alpha)/f_\eta(\alpha) = 1$ for all α. Hence, the theorem follows.   □

To reiterate, differential entropies are unlike absolute entropy in that differential entropy is not always positive and not necessarily finite. Another property is that differential entropy is not invariant to a one-to-one transformation of the random variable. We show this last fact in the following.

Given an absolutely continuous random variable X with pdf $f_X(x)$ and the transformation $Y = aX + b$, find an expression for the entropy of Y in terms of $h(X)$. How has the transformation affected the result?

Given $f_X(x)$ and $Y = aX + b$, then we have,

$$f_Y(y) = \frac{f_X(x)}{|a|} \bigg|_{x = \frac{y-b}{a}}$$

$$= \frac{f_X\left(\frac{y-b}{a}\right)}{|a|} .$$

Continuing from Eq. (2.24),

$$h(Y) = - \int_{-\infty}^{\infty} \frac{f_X\left(\frac{y-b}{a}\right)}{|a|} \log \frac{f_X\left(\frac{y-b}{a}\right)}{|a|} dy$$

$$= - \int_{-\infty}^{\infty} \frac{f_X\left(\frac{y-b}{a}\right)}{|a|} \left\{ \log f_X\left(\frac{y-b}{a}\right) - \log |a| \right\} dy .$$

Now,

$$\int_{-\infty}^{\infty} f_Y(y) \, dy = \int_{-\infty}^{\infty} \frac{f_X\left(\frac{y-b}{a}\right)}{|a|} dy = 1$$

and $x = \frac{y-b}{a}$, hence

$$- \int_{-\infty}^{\infty} \frac{f_X\left(\frac{y-b}{a}\right)}{|a|} \log f_X\left(\frac{y-b}{a}\right) dy = h(X) ,$$

so

$$h(Y) = h(X) + \log |a| .$$

The transformation has changed the differential entropy by an additive constant.

Example 2.14 Consider the discrete random variable X with alphabet $\mathcal{X} = \{x_1, x_2, x_3, x_4\}$ each with their respective probability $P_X(x_1) = \tfrac{1}{2}$, $P_X(x_2) = \tfrac{1}{4}$, $P_X(x_3) = P_X(x_4) = \tfrac{1}{8}$, and the linear transformation $Y = aX + b$. Find $H(Y)$. What effect has the transformation had on the entropy of the discrete random variable X? The probability density function for Y is

$$P_Y(y_1) = P_Y(ax_1 + b) = \frac{1}{2},$$

$$P_Y(y_2) = P_Y(ax_2 + b) = \frac{1}{4},$$

$$P_Y(y_3) = P_Y(ax_3 + b) = \frac{1}{8},$$

$$P_Y(y_4) = P_Y(ax_4 + b) = \frac{1}{8}.$$

Thus, from Eq. (2.2),

$$H(Y) = -\sum_{j=1}^{4} P_Y(y_j) \log P_Y(y_j)$$

$$= -\sum_{j=1}^{4} P_Y(ax_j + b) \log P_Y(ax_j + b).$$

But $P_Y(ax_j + b) = P_X(x_j)$, so

$$H(Y) = -\sum_{j=1}^{4} P_X(x_j) \log P_X(x_j) = H(X).$$

The transformation has not changed the discrete entropy.

Example 2.15   For a general multivariate probability density function of absolutely continuous random variables denoted by $f_X(x_1, x_2, \ldots, x_N)$, consider the one-to-one transformation represented by $y_i = g_i(X), i = 1, 2, \ldots, N$. We find an expression for the differential entropy of the joint pdf of Y, $f_Y(y_1, y_2, \ldots, y_N)$ in terms of the Jacobian of the transformation,

$$f_X(x_1, x_2, \ldots, x_N), \quad \text{and} \quad y_i = g_i(X), so$$

$$f_Y(y_1, y_2, \ldots, y_N) = \left. \frac{f_X(x_1, x_2, \ldots, x_N)}{|J(x_1, x_2, \ldots, x_N)|} \right|_{x_j = x_{ji}}.$$

Assuming that there is only one solution to $y_i = g_i(X)$, then from Eq. (2.24),

$$h(Y) = E\{-\log f_Y(y_1, \ldots y_N)\}$$
$$= E\{-\log f_X(x_1, x_2, \ldots, x_N)\} + E\{\log |J|\}$$
$$= h(X) + E\{\log |J|\},$$

where $J = J(x_1, x_2, \ldots, x_N)$.

Thus, the differential entropies differ by the expected value of the logarithm of the Jacobian.

The mutual information of two jointly distributed continuous random variables, say W and X, can also be defined as

$$I(W; X) = \int_{-\infty}^{\infty} \int_{-\infty}^{\infty} f_{WX}(w, x) \log \frac{f_{WX}(w, x)}{f_W(w) f_X(x)} \, dw \, dx$$
$$= I(X; W) \,. \tag{2.35}$$

As in the discrete case, the mutual information can be expressed in terms of differential (here) entropies as

$$I(W; X) = h(W) - h(W|X)$$
$$= h(X) - h(X|W) \,, \tag{2.36}$$

where

$$h(W|X) = - \int_{-\infty}^{\infty} \int_{-\infty}^{\infty} f_{WX}(w, x) \log f_{W|X}(w|x) \, dw \, dx \,. \tag{2.37}$$

Fortunately for our subsequent uses of $I(W; X)$, the mutual information is invariant under any one-to-one transformation of the variables, even though the individual differential entropies are not. We illustrate this by the following example.

Example 2.16   Show that mutual information is invariant to one-to-one transformations. That is, given two continuous random vectors X and Z with mutual information $I(X; Z)$ and a one-to-one transformation $y_i = g_i(X), i = 1, 2, \ldots, N$, find $I(Y; Z)$. >Straightforwardly, we know that

$$I(X; Z) = h(X) - h(X|Z)$$

and

$$I(Y; Z) = h(Y) - h(Y|Z) \,.$$

We are given that Y is a one-to-one transformation of X, so from a prior result

$$h(Y) = h(X) + E\{\log|J|\}$$

and

$$h(Y|Z) = h(X|Z) + E\{\log|J|\} \,.$$

Therefore,

$$I(Y; Z) = h(Y) - h(Y|Z)$$
$$= h(X) - h(X|Z) = I(X; Z) \,.$$

The mutual information is invariant under a one-to-one transformation since the term involving the Jacobian is in both differential entropy expressions and subtracts out.

Example 2.17   In this example we compute the differential entropy of the input and the mutual information between the input and output of an additive Gaussian noise channel with zero mean and variance σ_c^2. The input is also assumed to be Gaussian with zero mean and variance σ_s^2. Representing the channel as shown in Fig. 2.4, we thus have that

$$f_W(w) = \frac{1}{\sqrt{2\pi}\sigma_s} e^{-w^2/2\sigma_s^2} \tag{2.38}$$

and

$$f_{X|W}(x|w) = \frac{1}{\sqrt{2\pi}\sigma_c} e^{-(x-w)^2/2\sigma_c^2} . \tag{2.39}$$

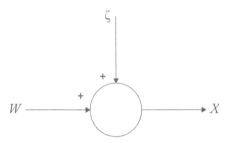

Figure 2.4: Additive noise channel.

The differential entropy of the input is

$$h(W) = -\int_{-\infty}^{\infty} f_W(w) \log f_W(w)\, dw \tag{2.40}$$

To calculate the mutual information, we choose to employ the expression $I(W; X) = h(X) - h(X|W)$. We already have $f_{X|W}(x|w)$, but we need $f_X(x)$. This follows directly as

$$f_X(x) = \frac{1}{\sqrt{2\pi}\left[\sigma_s^2 + \sigma_c^2\right]^{1/2}} e^{-x^2/2(\sigma_s^2 + \sigma_c^2)} , \tag{2.41}$$

since the input and the noise are independent, zero-mean Gaussian processes. By analogy with Eq. (2.40), we have from Eqs. (2.39) and (2.41) that

$$h(X) = \tfrac{1}{2} \log 2\pi e\left[\sigma_s^2 + \sigma_c^2\right] \tag{2.42}$$

and

$$h(X|W) = \tfrac{1}{2} \log 2\pi e \sigma_c^2 \ . \tag{2.43}$$

We thus find the mutual information to be

$$I(W; X) = \tfrac{1}{2} \log 2\pi e \left[\sigma_s^2 + \sigma_c^2\right] - \tfrac{1}{2} \log 2\pi e \sigma_c^2$$
$$= \tfrac{1}{2} \log \left(1 + \frac{\sigma_s^2}{\sigma_c^2}\right) \ . \tag{2.44}$$

In Eq. (2.44), we see that for $\sigma_s^2 \ll \sigma_c^2$, $I(W; X) \cong 0$, while if $\sigma_c^2 \to 0$, then $I(W; X) \to \infty$.

2.4 RELATIVE ENTROPY AND MUTUAL INFORMATION

The entropy of a random variable is a measure of the uncertainty; it is a measure of the information required on the average to describe the random variable [5, 6]. We have seen how entropy relates to mutual information, and now we consider the concept of relative entropy.

The relative entropy is a measure of the distance between two distributions. It is a measure of the inefficiency of assuming that the distribution is q when the true distribution is p. For example, if we knew the distribution, we could construct a lossless source code based on this knowledge (and we will do so in the following chapter). However, if instead of the true distribution p, we designed the code for a distribution q, what would be the resulting inefficiency in coding a random variable with distribution p? We will see in the next chapter that we would need an additional $D(p \parallel q)$ bits on the average above the minimum possible to describe the random variable. This quantity, $D(p \parallel q)$, called the relative entropy helps us represent this and other inefficiencies due to our incorrect knowledge of nature.

Definition 2.18 The *relative entropy* or Kullback Leibler distance between two probability mass functions $p(x)$ and $q(x)$ is defined as

$$D(p \parallel q) = \sum_x p(x) \log \frac{p(x)}{q(x)} \tag{2.45}$$
$$= E_p \log \frac{p(x)}{q(x)} \tag{2.46}$$

In this definition, we use the convention (based on continuity arguments) that $O \log \frac{o}{q} = 0$ and $p \log \frac{p}{o} = \infty$.

We can show that

$$D(p\|q) \geq 0 \tag{2.47}$$

with equality if and only if

$$p(x) = q(x) \quad \text{for all} \quad x \ . \tag{2.48}$$

This last result is called the Information Inequality and can be shown as follows.

Proof. We consider only the values of x such that $p(x) > 0$, called the support set of $p(x)$. Then

$$
\begin{aligned}
-D(p\|q) &= -\sum_{x \in A} p(x) \log \frac{p(x)}{q(x)} \\
&= \sum_{x \in A} p(x) \log \frac{q(x)}{p(x)} \\
&\leq \log \sum_{x \in A} p(x) \log \frac{q(x)}{p(x)} \leq \log \sum_{x \in X} q(x) \\
&= \log 1 = 0 \,.
\end{aligned}
\tag{2.49}
$$

Since $\log t$ is a strictly concave function of t, we have equality iff $q(x)/p(x) = 1$ everywhere. Hence we have $D(p\|q) = 0$ if and only if $p(x) = q(x)$ for all x. $\qquad\square$

We can use relative entropy to explore the properties of mutual information. For two random variables $(X, Y) \sim p(x, y)$ with marginals $p(x)$ and $p(y)$, the *mutual information* $I(X, Y)$ is the relative entropy between the joint distribution and the product distribution,

$$
I(X, Y) = \sum_x \sum_y p(x, y) \log \frac{p(x, y)}{p(x)p(y)}
\tag{2.50}
$$

$$
= D(p(x, y) \,\|\, p(x)p(y))
\tag{2.51}
$$

$$
= E_{p(x,y)} \log \frac{p(x, y)}{p(x)p(y)} \,.
\tag{2.52}
$$

The relative entropy leads to a direct proof of the non-negativity of mutual information. Explicitly, since $I(X;Y) = D(p(x, y) \,\|\, p(x)p(y)) \geq 0$ with equality if and only if $p(x, y) = p(x)p(y)$, i.e., X and Y are independent, then

$$
I(X;Y) \geq 0 \,,
\tag{2.53}
$$

with equality if and only if X and Y are independent.

The relative entropy can be used to show that the uniform distribution over the range X is the maximum entropy distribution over this range. Letting $u(x) = \frac{1}{|\mathcal{X}|}$ be the uniform probability mass function (pmf) over X and $p(x)$ be the actual probability mass function for X, then we have directly that

$$
\begin{aligned}
D(p\|u) &= \sum p(x) \log \frac{p(x)}{u(x)} \\
&= \log |\mathcal{X}| - H(X) \,.
\end{aligned}
\tag{2.54}
$$

but, we know that $D(p\|u) \geq 0$, so

$$
\log |\mathcal{X}| \geq H(X) \,.
$$

This inequality is of fundamental importance in lossless source coding.

2.5   DATA PROCESSING INEQUALITY

When analyzing communications systems and the inherent signal processing using information theory, an extremely useful result is the Data Processing Inequality [5]. The Data Processing Inequality is a consequence of the Markov chain relationships that often exist in communication systems. We begin with the definition of a Markov chain, which is, not surprisingly, important in its own right.

Definition 2.19   Random variables X, Y, Z are said to form a Markov Chain in that order (denoted $X \rightarrow Y \rightarrow Z$) if

$$p(x, y, z) = p(x)p(y|x)p(z|y) . \tag{2.55}$$

This simple result has several immediate consequences. One such consequence is that $X \rightarrow Y \rightarrow Z$ if and only if X and Z are conditionally independent given Y. The Markov chain assumption yields conditional independence since

$$
\begin{aligned}
p(x, z|y) &= \frac{p(x, y, z)}{p(y)} = \frac{p(x, y)p(z|x, y)}{p(y)} \\
&= \frac{p(x, y)p(z|y)}{p(y)} = p(x|y)p(z|y) .
\end{aligned}
\tag{2.56}
$$

Another consequence is that $X \rightarrow Y \rightarrow Z$ implies that $Z \rightarrow Y \rightarrow X$, which can also be expressed as $X \leftrightarrow Y \leftrightarrow Z$. We see that this is true by expanding the joint probability expression and manipulating the conditional marginals as

$$
\begin{aligned}
p(x, y, z) &= p(x)p(y|x)p(z|y) \\
&= p(x|y)p(y)p(z|y) \\
&= p(x|y)p(y|z)p(z) .
\end{aligned}
$$

Finally, it is relatively obvious that if $Z = f(Y)$, then $X \rightarrow Y \rightarrow Z$.

These results set up the following.

Theorem 2.20   (Data Processing Inequality).   *If $X \rightarrow Y \rightarrow Z$, then $I(X; Y) \geq I(X; Z)$.*

Proof. To show this, we use the chain rule to write mutual information in two different ways,

$$
\begin{aligned}
I(X; Y, Z) &= I(X; Z) + I(X; Y|Z) \tag{2.57} \\
&= I(X; Y) + I(X; Z|Y) . \tag{2.58}
\end{aligned}
$$

Evaluating the last term first, $I(X; Z|Y) = 0$ since X and Z are conditionally independent given Y. Further, noting that $I(X; Y|Z) \geq 0$, we have

$$I(X; Y) \geq I(X; Z) \tag{2.59}$$

with equality if and only if $I(X; Y|Z) = 0$, i.e., $X \to Z \to Y$ forms a Markov chain. Similarly, it can be shown that $I(Y; Z) \geq I(X; Z)$.

In many applications, we may measure Y and use signal processing to obtain $Z = g(Y)$. Is this a good thing to do? The following result provides an interesting insight.

If $Z = g(Y)$, then $I(X; Y) \geq I(X; g(Y))$. This follows directly from the fact that $X \to Y \to g(Y)$ forms a Markov chain. Hence, processing the data Y cannot increase the information about X! Why do we do this then?

We now present two additional results concerning relative entropy. The conditional relative entropy $D(p(y|x) \| q(y|x))$ is

$$D(p(y|x) \| q(y|x)) = \sum_x p(x) \sum_y p(y|x) \log \frac{p(y|x)}{q(y|x)} . \tag{2.60}$$

$$= E_{p(x,y)} \log \frac{p(Y|X)}{q(Y|X)} . \tag{2.61}$$

The notation for conditional relative entropy does not mention the distribution $p(x)$ of the conditioning random variable and thus does not explicitly indicate the averaging involved.

Finally, and perhaps not surprisingly, the relative entropy between two joint distributions for a pair of random variables can be expanded as the sum of a relative entropy and a conditional relative entropy,

$$D(p(x, y) \| q(x, y)) = \sum_x \sum_y p(x, y) \log \frac{p(x, y)}{q(x, y)}$$

$$= \sum_x \sum_y p(x, y) \log \frac{p(y|x)p(x)}{q(y|x)q(x)}$$

$$= \sum_x \sum_y p(x, y) \log \frac{p(x)}{q(x)} + \sum_x \sum_y p(x, y) \log \frac{p(y|x)}{q(y|x)}$$

$$= D(p(x) \| q(x)) + D(p(y|x) \| q(y|x)) . \tag{2.62}$$

$\square$

2.6   NOTES AND ADDITIONAL REFERENCES

The topical coverage in this chapter is relatively standard, although clearly selective compared to other books on information theory and rate distortion theory. Beyond the key concepts of entropy and mutual information, the most useful topics for later analyses are the chain rules, relative entropy, and the Data Processing Inequality.

CHAPTER 3

Lossless Source Coding

3.1 THE LOSSLESS SOURCE CODING PROBLEM

We now begin to interpret the quantities developed in Chapter 2 within a communications context [4, 5]. For this particular section, we consider the communication system block diagram in Fig. 1.2 under the assumptions that the channel is ideal (no noise or deterministic distortion) and the channel encoder/decoder blocks are identity mappings (a straight wire connection from input to output of each of these blocks). We are left with a communication system block diagram consisting of a source, source encoder, ideal channel, source decoder, and user. Since the channel is ideal and the channel encoder and decoder are identities, we have, a block diagram of the simplified communication system as shown in Fig. 3.1.

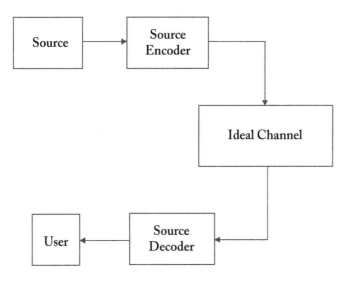

Figure 3.1: Simplified block diagram for the source coding problem.

Just what are the physical meanings of each of the components in Fig. 3.1? The source is some kind of data generation device, such as a computer or computer peripheral, which generates a discrete-valued random variable U with alphabet $\mathcal{U} = \{1, 2, \ldots, M\}$ and probability assignment $P_U(\cdot)$. For the development in this section, we assume that the successive letters produced by the source are statistically independent. Such a source is called a *discrete memoryless source* (DMS).

The ideal channel can be thought of as some perfectly operating communications channel or a mass storage device that is error-free. The user is a machine or individual that requires the data to accomplish a task. With these descriptions thus far, it is unclear why we need the source encoder/decoder blocks or what utility they might be. The answer is that data at the output of any given discrete source may not be in a form that yields the minimum required transmitted bit rate. The following example illustrates this point.

Example 3.1 [23]   The source output is a ternary-valued random variable that takes on the values $\{1, 2, 3\}$ with probabilities $P(1) = 0.7$, $P(2) = 0.15 = P(3)$. The source letter produced at each time instant is assumed to be independent of the letter produced at any other time instant, so that we have a DMS. We wish to find a binary representation of any sequence of ternary source letters such that the source sequence can be recovered exactly (by the decoder) and such that the average number of binary digits per source letter is a minimum.

A straightforward assignment of binary words, called a source code, is to let $u = 1$ be represented by 00, $u = 2$ by 10, and $u = 3$ by 11. This code transmits an average of 2 bits per source letter. Since $u = 1$ occurs much more often than the other two letters, it seems intuitive that we should use a shorter sequence to represent $u = 1$ than those used for $u = 2$ and $u = 3$. One such code assignment is

$$u = 1 \rightarrow 0$$
$$u = 2 \rightarrow 10$$
$$u = 3 \rightarrow 11 \ .$$

By inspection, it is clear in this simple example that each symbol of the original ternary source data can be uniquely recovered from its binary codeword, so we say that the code is *uniquely decodable*. Of course, this is a minimum requirement of a code. This code has an even more desirable property that no codeword is a prefix of any other codeword, called the *prefix condition*, so as soon as a codeword appears, it can be decoded. There is no need to wait for a longer string of symbols to start decoding.

The average number of bits required per source letter, denoted here by $\bar{l}$, is thus

$$\bar{l} = 1 \cdot P_U(1) + 2 \cdot P_U(2) + 2 \cdot P_U(3)$$
$$= 0.7 + 0.3 + 0.3 = 1.3 \text{ bits/source letter} \ . \tag{3.1}$$

This is a clear improvement over the original 2-bits/source letter code.

To try and reduce $\bar{l}$ further, we encode pairs of source letters, which are listed in the accompanying table together with the probability of each pair. If we now assign a binary word to each pair of source letters as shown in the column labeled "codeword" in the table, we find that the average binary codeword length per source letter is

$$\bar{l} = \tfrac{1}{2}\{1(0.49) + 3(0.105) + 3(0.105) + 3(0.105) + 4(0.105) + 6(0.0225)$$
$$+ 6(0.0225) + 6(0.0225) + 6(0.0225)\}$$
$$= 1.1975 \text{ bits/source letter} \ . \tag{3.2}$$

Table 3.1: Pairs of Ternary Letters and a Source
Code for Ex. 3.1

Source Letters	Probability	Codeword
u_1u_1	0.49	0
u_1u_2	0.105	111
u_1u_3	0.105	100
u_2u_1	0.105	101
u_3u_1	0.105	1100
u_2u_2	0.0225	110110
u_2u_3	0.0225	110111
u_3u_2	0.0225	110100
u_3u_3	0.0225	110101

Source: A. D. Wyner, "Fundamental Limits in
Information Theory," *Proc. IEEE,* © 1981 IEEE.

The value in Eq. (3.2) is slightly better than the 1.3 bits/letter achieved by our second code. The code in the table is also uniquely decodable back into the original ternary sequence and satisfies the prefix condition so that it is instantaneously decodable.

Although we have not described how the binary codes were selected, it is clear that at least in this particular case, source coding allows the ternary data to be represented with a smaller number of bits than one might originally think. Thus, the utility of the source encoder/decoder blocks in our basic system block diagram is demonstrated.

3.2 DEFINITIONS, PROPERTIES, AND THE SOURCE CODING THEOREM

Now that we have seen an example of a lossless source code and we see the efficiency gained by such a code, we now must determine ways to design such codes [4, 5]. In order to do this, it is also helpful to observe the desirable properties of the example source code and specify any other properties that we might desire. Certainly we will want our source codes to be uniquely decodable and the prefix condition is also a very nice property too, since it makes decoding so easy. Some other conditions might be that the design process should be simple and that the code should be efficient.

This last point is somewhat more subtle. How do we determine if a code is efficient? In the example given, we showed improvement over the two bit per symbol codeword assignment in that the *average* codeword length of the code was reduced. We also demonstrated that by creating pairwise blocks of source symbols, the average codeword length could be reduced further. Thus, we have chosen minimizing the average codeword length as our measure of efficiency. We make these ideas more formal with a couple of definitions.

Definition 3.2   A *source code C* for a random variable X is a mapping from $\mathcal{X}$, the range of X, to D^*, the set of finite length strings of symbols from a D-ary alphabet. Let $c(x)$ denote the codeword corresponding to x and let $l(x)$ denote the length of each $c(x)$.

Definition 3.3   The expected length of a source code $C(x)$ for a random variable X with pmf $p(x)$ is given by

$$\bar{l}(C) = \sum_{x \in \mathcal{X}} p(x) l(x) \tag{3.3}$$

where $l(x)$ is the length of the codeword associated with x.

A code is uniquely decodable if any string of codewords has only one possible string of source words that can produce it. Of course, in order to decode the string, it may be necessary to look at the entire string of codewords to determine the unique string of source symbols.

A code is a *prefix code* or an *instantaneously decodable code* if no codeword is a prefix of any other codeword. This means that a codeword can be decoded as soon as a full codeword occurs. There is no need to look ahead at later codewords.

In the example in the previous section, codes were exhibited that produced a reduction in the average codeword length. Could we have kept going and done even better? In particular, what is the lowest rate (average codeword length) achievable by any lossless source code design? The answer to this question is available from Shannon's first theorem, the Source Coding Theorem, which can be stated as follows.

Theorem 3.4   (Source Coding Theorem).   *For a DMS with entropy $H(U)$, the minimum average codeword length per source letter ($\bar{l}$) for any code is lower bounded by $H(U)$, that is, $\bar{l} \geq H(U)$, and further, $\bar{l}$ can be made as close to $H(U)$ as desired for some suitably chosen code.*

A proof is given at the end of this chapter when we develop the Asymptotic Equipartition Property. We give a motivation here for the importance of the entropy $H(U)$ and how the average codeword length $\bar{l}$ is related to the entropy.

We can write the difference between the expected length and entropy as

$$\bar{l} - H_D(X) = \sum_i p_i l_i - \sum_i p_i \log_D \frac{1}{p_i} \tag{3.4}$$

$$= -\sum_i p_i \log_D D^{-l_i} - \sum_i p_i \log_D \frac{1}{p_i} . \tag{3.5}$$

Letting

$$r_i = \frac{D^{-l_i}}{\sum_j D^{-l_j}} \quad \text{and} \quad C = \sum D^{-l_i} \,,$$

we obtain

$$\bar{l} - H = \sum_i p_i \log_D \frac{p_i}{r_i} - \log_D C \tag{3.6}$$

$$= D(p\|r) + \log_D \left(\frac{1}{C}\right) \geq 0$$

since $D(p\|r) \geq 0$ and $C \leq 1$. Hence, $L \geq H$ with equality iff $p_i = D^{-l_i}$, i.e., iff $-\log_D p_i$ is an integer for all i.

Therefore, the (absolute) entropy of a discrete source is thus a very important physical quantity, since it specifies the minimum bit rate required to yield a perfect replication of the original source sequence. Revisiting the prior example, we see that for the DMS in Ex. 3.1, since

$$H(U) = -0.7 \log 0.7 - 0.15 \log 0.15 - 0.15 \log 0.15$$
$$= 1.18129\ldots \,, \tag{3.7}$$

we know that $\bar{l} \geq 1.18129\ldots$, and hence only a slight further reduction in rate (from 1.1975 bits/source letter) can be achieved by designing additional source codes. So, this may be a good place to stop.

3.3   HUFFMAN CODING AND CODE TREES

Theorem 3.4 states that a lower bound on the average codeword length per source letter, denoted by $\bar{l}$, is the source entropy [4, 5]. Furthermore, Example 2.2 illustrates that a DMS with letters that are not equally likely has a smaller entropy than a source with the same number of letters and equally probable outputs, and in Ex. 3.1, a variable-length code is constructed for a DMS with nonequally likely output letters.

The design of variable-length codes with an $\bar{l}$ that approaches the entropy of the DMS is generically referred to as *entropy coding*, and there are several procedures for finding such codes. In this section we present the most familiar and most straightforward of the available techniques for entropy coding, which is due to Huffman [24] and is thus called *Huffman coding*. To specify the encoding procedure, we consider a DMS U with M output letters $\{1, 2, \ldots, M\}$ and probabilities $P_U(u), u = 1, 2, \ldots, M$. We also assume for simplicity that the letters are numbered such that $P_U(1) \geq P_U(2) \geq \cdots \geq P_U(M)$. Of course, if this property does not hold at the outset, we can always renumber the letters to produce it.

The constructive procedure for designing the variable-length code can be described as follows. The letters and their probabilities are listed in two columns in the order of decreasing probability. The two lowest-probability letters are combined by drawing a straight line out from each and connecting them. The probabilities of these two letters are then summed, and this sum is considered to be the probability of a new letter denoted $(M-1)'$. The next two lowest-probability letters, among $1, 2, \ldots, M-2$, and $(M-1)'$, are combined to create another letter with probability equal to their sum. This process is continued until only two letters remain and a type of "tree" is generated. Binary codewords are then assigned by moving from right to left in the tree, assigning a 0 to the upper branch and a 1 to the lower branch, where each pair of letters has been combined. The codeword for each letter is read off the tree from right to left. An example is given to clarify the procedure.

Example 3.5   Given a DMS U with five letters and probabilities $P_U(1) = 0.5$, $P_U(2) = P_U(3) = 0.2$, and $P_U(4) = P_U(5) = 0.05$, we wish to design a binary variable-length code for this source. With reference to Fig. 3.2, we combine the two least likely letters and sum their probabilities to create a new "letter."

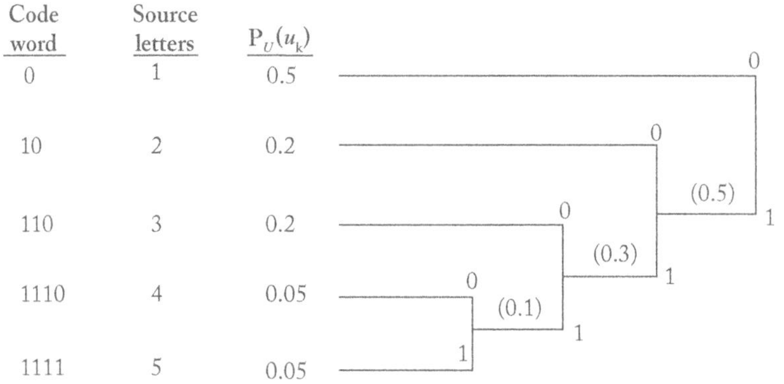

Figure 3.2: Huffman encoding for the DMS in Ex. 3.5.

The least likely letters are combined again, and the procedure is continued until only two letters remain. Codewords are then assigned by moving right to left and assigning a 0 to an upper branch and a 1 to a lower branch. Codewords are also read off in a right-to-left fashion and are shown in the leftmost column of Fig. 3.2. The average codeword length $(\bar{l})$ for this variable-length code is

$$\bar{l} = (4)(0.05) + (4)(0.05) + (3)(0.2) + (2)(0.2) + (1)(0.5)$$
$$= 1.9 \text{ bits/source letter .} \tag{3.8}$$

We know that $H(U)$ is upper bounded by $\log_2 M = \log_2 5 \cong 2.322$ bits/source letter, and we can show that the entropy of U is $H(U) \cong 1.86$ bits/source letter. Thus, $H(U) < \bar{l} < \log_2 M$.

Example 3.6   Given a DMS U with four letters and probabilities $P_U(1) = 0.5$, $P_U(2) = 0.25$, and $P_U(3) = P_U(4) = 0.125$, use the Huffman procedure to design a variable-length code. Find $\bar{l}$ and compare to $H(U)$. The procedure is shown in Fig. 3.3 and the corresponding average codeword length is

$$\bar{l} = 1(0.5) + 2(0.25) + 3(0.125) + 3(0.125) = 1.75 \text{ bits/source letter}$$

We can calculate the discrete source entropy as

$$\begin{aligned} H(U) &= -0.5 \log_2 0.5 - 0.25 \log_2 0.25 - 0.125 \log_2 0.125 - 0.125 \log_2 0.125 \\ &= 0.5 + 0.5 + 0.375 + 0.375 \\ &= 1.75 \text{ bits/source letter .} \end{aligned}$$

Therefore, we conclude that we cannot do any better with any code since $\bar{l} = H(U)$.

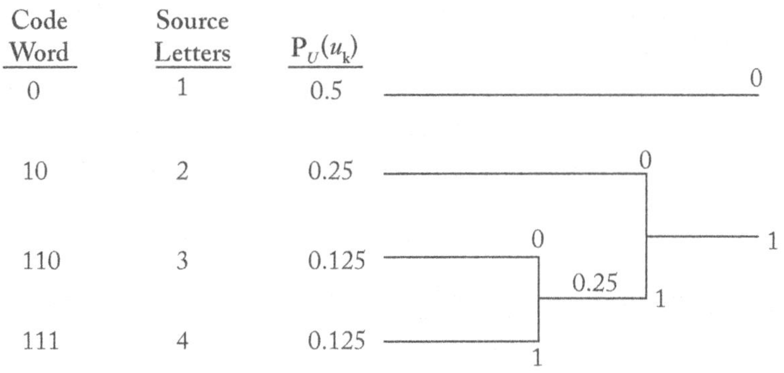

Code Word	Source Letters	$P_U(u_k)$
0	1	0.5
10	2	0.25
110	3	0.125
111	4	0.125

Figure 3.3: Huffman Coding for Example 3.6

Is it possible to know in advance, without forming the Huffman code and calculating the source entropy, that a source will yield an average codeword length exactly equal to the entropy? The answer is yes. If the probabilities of the source letters are all negative powers of the code alphabet, say negative powers of 2 in the case of a binary code, the source is called *diadic* and the average codeword length will equal the entropy. We can confirm the diadic property for the source in the last example, and we also see that the probabilities of the source letters for the other sources in prior Huffman coding examples are not diadic and that the Huffman procedure does not yield an average codeword length equal to the entropy of the source.

The codewords in Ex. 3.5 are *uniquely decodable* in that each source letter has a codeword that differs from that assigned to any other letter. *The Huffman procedure yields the smallest average codeword length of any uniquely decodable set of codewords.* Although another code can do as well as the Huffman code, none can be better.

It is important to recognize that Huffman codes can be represented by a code tree. Of course, this is evident from the basic construction and the assignment of the codewords from right-to-left in the tree. The tree structure is in fact very critical in analyzing and understanding prefix condition codes. Specifically, in a binary code tree where at each node 0 is assigned to one branch and a 1 is assigned to the other branch, one can start with a full tree, meaning all branches present to some depth, and prune the tree to obtain a prefix condition code.

Considering the most recent Huffman coding example, where the resulting codewords were 0, 10, 110, and 111, we see that if we started with a full code tree to depth 3, and pruned all branches emanating from the first branch labeled 0, and then at the next step, prune all branches emanating form the 10 branch, and then left the last two branches, we get this code exactly. Further, by pruning all branches emanating from a node, we invoke the prefix condition. Code trees are useful in representing, analyzing, and understanding many lossless coding approaches.

Recall that in Ex. 3.1, a code is designed for single-letter encoding of the source, and then another code is designed to represent pairs of source letters. Should we use Huffman coding on pairs of letters for the source in Ex. 3.5? Probably not, since $\bar{l}$ is already very close to the entropy. However, Ex. 3.1 demonstrates that by encoding blocks of source letters, an average codeword length nearer the source entropy may be obtained (with added complexity). A more general illustration of this property is provided by the following two inequalities. The Huffman encoding procedure generates a code for the DMS U with an $\bar{l}$ that satisfies

$$H(U) \leq \bar{l} < H(U) + 1 \text{ bits/source letter} \tag{3.9}$$

for letter-by-letter (or symbol-by-symbol) encoding. If blocks of L letters are combined before using the Huffman technique, $\bar{n}$ is bounded by

$$H(U) \leq \bar{l} < H(U) + \frac{1}{L} \text{ bits/source letter} . \tag{3.10}$$

Thus, encoding pairs of letters is at least as good as single-letter encoding, and for large L, $\bar{l}$ can be made arbitrarily close to $H(U)$. Whether block encoding makes sense depends on the particular DMS and its letter probabilities (and perhaps, the application of interest).

Only the binary Huffman procedure has been described here, but nonbinary codes can be designed using the Huffman method. The details are slightly more complicated and nonbinary codes are less commonly encountered than binary ones, so further developments are left to the references [18, 22].

We see that one way to potentially improve the performance of Huffman codes is to form blocks of source symbols and design Huffman codes for these blocks. However, when the source alphabet size is small and the probabilities of the source letters is highly skewed, the number of symbols in a block may need to be more than a few, thus leading to a large source alphabet size and a large number of codewords. This complicates the design process and the storage and implementation complexity of the code grows accordingly. Further, some of the source symbols created by blocking may have a very low probability of occurrence, and in reality, codewords for only a few of these symbols may ever be needed.

Another requirement of the basic Huffman coding technique presented thus far is that the probabilities of the individual source letters must be known. Huffman codes can still be constructed "on the fly" in these situations by growing trees that adaptively learn the relative frequencies of occurrence of the letters as they appear. This adaptive Huffman coding method is not presented here but it is described in detail in Sayood [18].

3.4   ELIAS CODING AND ARITHMETIC CODING

As an alternative to Huffman coding, it would be useful to have a way to code individual sequences as they appear without generating the entire codebook [5, 18]. Arithmetic coding is such a lossless coding technique. A predecessor to arithmetic coding, which demonstrates the idea, although it is impractical to implement directly, is what is usually called Elias coding. For simplicity we treat the simple, but illustrative case of a binary source.

Generally, Elias coding treats an incoming binary sequence as a number between 0 and 1 by inserting a "decimal" point or "binary" point in front of the sequence of binary digits. Interpreting the resulting binary number as a sum of negative powers of 2, we get a number in the interval [0,1]. Thus, the resulting binary number can be associated with a subinterval in [0,1]. The concept of generating a code by aligning sequences with intervals in [0,1] may seem unusual at first blush, but the origin of this idea was in Shannon-Fano coding, which we now describe.

SHANNON-FANO CODING

Given the source alphabet $X = \{1, 2, \ldots, M\}$ all letters with $p(x) > 0$, the cumulative distribution function (CDF) is given by

$$F(x) = \sum_{\alpha \leq x} p(\alpha) \, .$$

In the following it will be useful to have the quantity

$$\bar{F}(x) = \sum_{\alpha < x} p(\alpha) + \frac{1}{2} p(x) \, .$$

We will use this quantity to develop a code (or tag) corresponding to the interval. In general, $\bar{F}(x)$ is a real number, so to use it as a code we must round it off to the codeword length $l(x)$ bits, which we denote as $\lfloor \bar{F}(x) \rfloor_{l(x)}$. This value falls within the corresponding interval and the intervals do not overlap, so a codeword can be based on this value. For a binary code alphabet, we express $\bar{F}(x)$ in binary with an appropriately placed "decimal" point and truncate to $l(x)$ bits.

Table 3.2: Example

x	$p(x)$	$F(x)$	$\bar{F}(x)$	$\bar{F}(x)$ in binary	$l(x)$	code
1	.25	.25	.125	0.001	3	001
2	.5	.75	.5	0.10	2	10
3	.125	.875	.8125	0.1101	4	1101
4	.125	1.0	.9375	0.1111	4	1111

This process is particularly simple for the diadic source in the last example of the prior section, where the source letter probabilities are as shown in the table below. Since the binary representation is exact, the choice of $l(x)$ for each codeword is clear, and the resulting codewords are shown in the table.

However, when the $\bar{F}(x)$ does not have a binary representation that naturally truncates, it is necessary to select a rule to specify $l(x)$. The rule for Shannon-Fano coding is

$$l(x) = \left\lceil \log \frac{1}{p(x)} \right\rceil + 1$$

bits.

As an example, we consider a source obtained by splitting the probability of 0.5 in the diadic example. For this source the code construction follows as shown in the following table.

Table 3.3: Example

$p(x)$	$F(x)$	$\bar{F}(x)$	$l(x)$	code
0.3	0.3	0.15	3	001
0.25	0.55	0.425	3	011
0.2	0.75	0.65	4	1010
0.125	0.875	0.8125	4	1101
0.125	1.0	0.9375	4	1111

If we examine the codes generated by this Shannon-Fano coding scheme, it is evident by inspection that there are some inefficiencies since we can prune bits off codewords and still have a prefix code. In fact, by definition of $l(x)$, we see that

$$\bar{l}(C) = \sum_x p(x)l(x) = \sum_x p(x)\left(\left\lceil \log \frac{1}{p(x)} \right\rceil + 1\right)$$
$$< H(X) + 2 .$$

Similar to Huffman coding, when we consider arithmetic coding for a sequence of letters, some L, then $\bar{l}$ is bounded by

$$H(U) \leq \bar{l} < H(U) + \frac{2}{L} \text{ bits/source letter .} \qquad (3.11)$$

In spite of this apparent inefficiency compared to Huffman coding, arithmetic codes allow the use of lossless coding without building a codebook, which can be very large for not-so-large blocklengths, and that can be a significant advantage for arithmetic codes. Further, when the source letter probabilities are not known, the arithmetic coding process lends itself more easily to separate source model estimation followed by coding than the adaptive Huffman procedure, which requires the growing and preservation of a tree.

Since it is desired to code sequences of binary symbols, a mapping is needed to map the incoming binary digit to the interval $[0,1]$. The cumulative distribution function is used for this purpose. The process is best understood by way of an example. We use Elias coding to demonstrate the concept.

GENERAL ITERATIVE RULE FOR ELIAS CODING OF BINARY SOURCES

Let $p = p_0 = 1 - p_1$ be the probability of a 0. After $n - 1$ source bits, an interval $[A_{n-1}, B_{n-1})$ has been specified. If the nth source symbol is a 0, the next bit specifies a new interval with endpoints

$$A_n = A_{n-1}$$
$$B_n = A_{n-1} + p(B_{n-1} - A_{n-1})$$

If the nth source symbol is a 1, the new interval is specified by the endpoints

$$A_n = A_{n-1} + p(B_{n-1} - A_{n-1})$$
$$B_n = B_{n-1}$$

For example, with $p = p_0 = 0.7$, $A_0 = 0$, $B_0 = 1$ and input 0110, we have the following.

0 in

$$A_1 = A_0 = 0$$
$$B_1 = 0 + 0.7(1 - 0) = 0.7$$

1 in

$$A_2 = A_1 + 0.7(B_1 - A_1) = 0 + 0.7(0.7 - 0) = 0.49$$
$$B_2 = B_1 = 0.7$$

1 in

$$A_3 = A_2 + 0.7(B_2 - A_2)$$
$$= 0.49 + 0.7(0.7 - 0.49) = 0.637$$
$$B_3 = B_2 = 0.7 \qquad \text{in } [0.625, 0.75) \quad \text{Release } 101$$

0 in

$$A_4 = A_3 = 0.637$$
$$B_4 = A_3 + 0.7(B_3 - A_3) \qquad \text{in } [0.625, 0.6875) \quad \text{Release } 0$$
$$= 0.637 + 0.7(0.7 - 0.637) = 0.6811$$

The released symbols are the generated codeword and so the code would be between 0.625 and 0.6875 or 1010.

The Elias coding scheme just demonstrated illustrates the basic concept behind arithmetic coding, and the example also reveals a clear challenge for implementation. Specifically, the method requires an ever-growing level of precision in the computations. Secondly, and less obvious, but still evident in the example, code symbols are generated quite asynchronously and unpredictably. In fact, it is not difficult to construct sources and examples where no code symbols are ever produced by the method! Berger presents some examples and insightful analyses of the Elias coding method in Chapter 3 of [22].

Practical arithmetic coding schemes had to wait for the work of Rissanen and Langdon in 1979 [25]. Sayood presents a very detailed and accessible development of arithmetic coding methods that avoid the infinite precision problem and that also address the retention problem [18]. Arithmetic coding is not developed further here.

One of the properties often cited for arithmetic coding is that it involves two separate steps–modeling and coding. The modeling step gives arithmetic coding a claim for universality and the coding step is designed to avoid the limitations of Elias coding.

There are many different approaches to lossless source coding and a set of lossless coding methods that are useful for the situation where it is only known that the source is stationary and ergodic are Lempel-Ziv coding schemes. These techniques fall in the class of universal coding methods and are introduced in the following section.

3.5   LEMPEL-ZIV CODING

The genesis of Lempel-Ziv coding comes from two papers by Jacob Ziv and Abraham Lempel published in 1977 and 1978 [27, 28], and the corresponding codes are denoted respectively as LZ77 and LZ78 [18, 26]. The LZ77 approach shifts a window of incoming symbols as they are coded, and the LZ78 approach creates a fixed length dictionary based on past and current input symbols. They both asymptotically approach the entropy of the source, although the LZ78 approach is more efficient for medium length sequences. A variation on the LZ78 method incorporates a modification due to Terry Welch that allows the efficient encoding of a special case that can occur. This modified algorithm is referred to as the LZW algorithm. We describe LZW encoding and decoding of binary sequences in the following.

LZW RULES FOR DICTIONARY ADDITIONS

- In the *decoder*, add the shortest string not present in the table with the previous output as the prefix and the letters of the next output as suffix [18, 26].

- At the *encoder*, find the longest matching sequence in the table and output its index. Add this sequence plus the next digit to the table.

THE WELCH MODIFICATION TO LZ78

The decoder needs a special rule when a string of the form $x\omega x\omega x$ occurs, when $x\omega$ already appears in the table [18, 26]. To see this, note that in this case, the encoder will send the index corresponding to $x\omega$ and add $x\omega x$ to the (encoder) table. The encoder will then see (parse) $x\omega x$, which appears in the encoder table, and send its index.

The decoder will receive the code for $x\omega x$ but it will not be in the decoder's table since it has not received an extension of $x\omega$. When this happens, that is, when an unknown code is received, the decoder can recognize this special case. In the special case, it knows that the code is for a string that has the previously decoded string as a prefix. Further, because of the form $x\omega x\omega x$, the decoder knows that the last digit is the same as the first digit of the previous decoded string (for example, prefix $\overbrace{?}^{x\omega} \Rightarrow x\omega x$).

To illustrate the encoding and decoding processes, consider an input sequence, 101100010101... . The encoding will proceed as shown in the following table, where we start the dictionary (codebook) with only the binary symbols 0 and 1 with their corresponding index 0 and 1, respectively. As we go through the encoding process, we find the longest string in the input sequence in the dictionary and output the corresponding index to be sent to the decoder. Following the encoding rule, we then add 10 to the encoder dictionary with index 2 to prepare for the next sequence of input symbols to be encoded. The Pointer is included in the table to keep track of the location of the next symbol to be encoded. The longest string in the input that is also in the dictionary is then found starting with the next input symbol, and this is found to be a 0 with an index of 0 in the dictionary. A 0 is then the output of the encoder and a new entry is added to the dictionary, which is now 01 with index 3. This process continues as shown in the encoding table.

The decoder starts with the same dictionary as the encoder, namely, symbols 0 and 1 with their respective index 0 and 1. The dictionary is grown as the decoding progresses. The decoder

Table 3.4: Encoder

Step	Longest string	Output	Add	Dictionary	Index	Pointer
				0	0	
				1	1	1
1	1	1	10	10	2	2
2	0	0	01	01	3	3
3	1	1	11	11	4	4
4	10	2	100	100	5	6
5	0	0	00	00	6	7
6	01	3	010	010	7	9
7	010	7	0101	0101	8	

receives the encoder output sequence, which consists of the dictionary index corresponding to the transmitted codeword. The decoder looks up the index in its dictionary and decodes the received symbol. The decoder then adds a symbol to its dictionary according to the rules provided, including the Welch modification. In the decoding table, note that when a symbol (index) is received at a certain step (numbered in the table), the content of the dictionary is specified by every entry prior to that step. Thus, when the decoder receives the symbol 7 at step 7, the dictionary only consists of entries 0 through 6 and the Welch rule must be invoked as shown.

Table 3.5: Decoder

Step	Received	Out	Add	Dictionary	Index
				0	0
				1	1
1	1	1	none	–	–
2	0	0	10	10	2
3	1	1	01	01	3
4	2	10	11	11	4
5	0	0	100	100	5
6	3	01	00	00	6
7	7	010	010	010	7
				0101	8

Of course, the LZW and related techniques will not be efficient when compressing short sequences, and there are decisions to be made in the algorithm implementation. Among these decisions is how to choose the size of the dictionary and how to prune the dictionary once the size limit has been reached. Two common pruning methods are to remove the earliest additions to the dictionary and to remove the entry that was used the longest time ago.

There are many variants on LZ coding and Toby Berber provides an excellent discussion and numerous example designs in Chapter 3 of Gibson, Berger, et al [22]. When analyzing LZ based codes, constructing code trees is very useful. Sayood shows how to construct both LZ77 and LZ78 codes with a clear step-by-step exposition. Both references are highly recommended for more details on Lempel-Ziv coding.

3.6 KRAFT INEQUALITY

When searching for good instantaneous or prefix condition codes, the Kraft Inequality is an important tool [4, 5]. In particular, the collection of codewords for any prefix condition or instantaneous code must have codeword lengths that satisfy the Kraft Inequality, which is stated and proved as follows for binary codes.

Theorem 3.7 (Kraft Inequality). *The codeword lengths $l_1, l_2, \ldots, l_M$ for any binary prefix condition code (instantaneous code) must satisfy the inequality*

$$\sum_i 2^{-l_i} \leq 1 . \tag{3.12}$$

Conversely, if a set of codeword lengths satisfies this inequality, there exists a prefix condition or instantaneous code with these codeword lengths.

The proof is instructive for further analyzing good codes so we present a version here.

Proof. Given a binary tree in which each node has 2 children, we let each branch of the tree represent a symbol of a codeword. Thus, at each node in the tree, two new branches are created, one branch for a 0 and the other branch for a 1. Traversing the tree from root node to the end of each path through the tree traces out a codeword.

In order to satisfy the prefix condition, no codeword is an ancestor of any other codeword on the tree; that is, each codeword prunes all possible descendants from the tree. Given l_{max}, the length of the longest codeword in the collection of codewords, a codeword at level l_i in the tree has $2^{l_{max}-l_i}$ descendants at level l_{max}. For codes that satisfy the prefix condition, each of the descendant sets must be disjoint. Additionally, the total number of terminal nodes for all of these disjoint sets must be less than or equal to $2^{l_{max}}$, which is the maximum number of branches in a tree to depth l_{max}. Summing over all the disjoint sets of all the codewords, we obtain

$$\sum_{i=1}^{M} 2^{l_{max}-l_i} \leq 2^{l_{max}} \tag{3.13}$$

or

$$\sum 2^{-l_i} \leq 1 \tag{3.14}$$

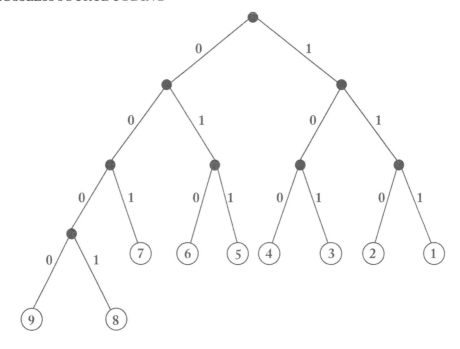

Figure 3.4: A four-level code tree for a binary code

which is the Kraft inequality for a binary code alphabet.

Conversely, for any set of codeword lengths, $l_1, l_2, \ldots, l_M$ that satisfy the Kraft inequality, we can always construct a prefix condition code with the specified codeword lengths, $l_1, l_2, \ldots, l_M$ by constructing a code tree. All we have to do is label the first node (lexicographically) of depth l_1, as codeword 1, remove its descendants from the tree, label the next remaining node of depth l_2 as codeword 2, and so on until we use all possible codeword lengths. □

The Kraft inequality holds for any D-ary alphabet and also for all uniquely decodable codes. See a four-level code tree for a binary code in Fig. 3.4.

The Kraft inequality can be used to find the prefix code with the minimum average codeword length by finding the set of lengths $l_1, l_2, \ldots, l_M$ that satisfy the Kraft inequality with minimum $\bar{l}(C)$. Specifically, we need to minimize

$$\bar{l} = \sum_i p_i l_i \tag{3.15}$$

over all integers satisfying

$$\sum 2^{-l_i} \leq 1 . \tag{3.16}$$

The proof is left to the references but we can take the idea one step further fairly directly.

Neglecting the integer constraint on the l_i and assuming equality in the (3.16) constraint, we use Lagrange multipliers to write

$$J = \sum_{p_i l_i} + \lambda \left(\sum D^{-l_i} \right) . \tag{3.17}$$

This, of course, yields the optimal codeword lengths

$$l_i^* = -\log_D p_i = \log_D \frac{1}{p_i} . \tag{3.18}$$

This possibly *non-integer* choice of codeword lengths yields expected codeword length

$$\bar{l}^* = \sum_i p_i l_i = -\sum p_i \log_D p_i = H_D(p_i) = H_D(X) , \tag{3.19}$$

since $\{p_i\}$ is the pmf for the source random variable X. But since the l_i must be integers, we will not always be able to set the codeword lengths as in (3.18). Instead, we should choose a set of codeword lengths "close" to the optimal set. Verifying that $l_i^* = -\log_D p_i$ is a global minimum is left as an exercise.

It is also possible to show that bounds on the average codeword length of an optimal code are the same as those we have already stated for Huffman codes. In particular, with $l_1^*, l_2^*, \ldots, l_M^*$ denoting the optimal codeword lengths for the source pmf p_i, then the optimal code satisfies

$$H_2(X) \le \bar{l}^* < H_2(X) + 1 . \tag{3.20}$$

The overhead in this result is due to the fact that $\log(\frac{1}{p_i})$ is not always an integer. As has been demonstrated in examples, we can reduce the overhead per symbol by spreading it out over many symbols.

To see this, consider a system which sends a sequence of n symbols from X, each drawn i.i.d. according to $p(x)$. Each set of n symbols is viewed as a supersymbol from $\mathcal{X}$. Define the expected codeword length per input symbol $\bar{l}_n$ as

$$\bar{l}_n = \frac{1}{n} \sum p(x_1, \ldots, x_n) l(x_1, \ldots, x_n) = \frac{1}{n} El(X_1, \ldots, X_n) . \tag{3.21}$$

We apply the previously derived bounds to this supersymbol code,

$$H(X_1, \ldots, X_n) \le El(X_1, \ldots, X_n) < H(X_1, \ldots, X_n) + 1 . \tag{3.22}$$

Since $X_1, X_2, \ldots, X_n$ are i.i.d.,

$$H(X_1, \ldots, X_n) = \sum H(X_i) = nH(X) ,$$

so using this in (3.22) and dividing by n,

$$H(X) \le \bar{l}_n < H(X) + \frac{1}{n} . \tag{3.23}$$

Thus, by using large block lengths we can achieve an expected codelength per symbol arbitrarily close to the entropy.

We can use the same argument for a sequence of not-necessarily i.i.d. samples from a stochastic process, and obtain similar bounds of the form

$$\frac{H(X_1, \ldots, X_n)}{n} \leq \bar{l}_n < \frac{H(X_1, \ldots, X_n)}{n} + \frac{1}{n} . \tag{3.24}$$

What happens to $\bar{l}_n$ if the code is designed for the wrong distance? Suppose that the code is designed for the pmf $q(x)$, so $l(x) = \lceil \log \frac{1}{q(x)} \rceil$, but the true the pmf is $p(x)$. Then, we have the result that (not Huffmann coding explicitly), the expected length under $p(x)$ of the code assignment $l(x) = \lceil \log \frac{1}{q(x)} \rceil$ satisfies

$$H(p) + D(p\|q) \leq E_p l(x) < H(p) + D(p\|q) + 1 . \tag{3.25}$$

The proof of both the upper and lower bounds involve adding and subtracting $\log p(x)$ and using the definition of relative entropy. Further details are left as an exercise.

3.7   THE AEP AND DATA COMPRESSION

A powerful tool for gaining insight into the role of entropy and asymptotics in data compression is the Asymptotic Equipartition Property or AEP [4, 5]. Shannon used a version of the AEP in his original paper in 1948 to demonstrate the importance of entropy for describing a source. Here we state a theorem that defines the AEP, and unusually for this book, we provide a proof. The proof is straightforward and does not require advanced mathematics, plus it allows us to introduce the idea of a typical set, which is key in more advanced developments of information theory and rate distortion theory.

Theorem 3.8   Asymptotic Equipartition Property (AEP).   *If $X_1, X_2, \ldots,$ are i.i.d. with pmf $p(x)$, then*

$$\frac{-1}{n} \log p(X_1, \ldots, X_n) \longrightarrow H(X) \quad \text{in probability} .$$

Proof.   Functions of independent random variables are also independent, so since the X_i are statistically independent, so are the $\log p(X_i)$. By Chebychev's inequality (see appendix)

$$P_r \left[\left| \frac{-1}{n} \log p(X_1, \ldots, X_n) - H(X) \right| \geq \varepsilon \right] \leq \frac{\text{var} \left[\frac{-1}{n} \log p(X_1, \ldots, X_n) \right]}{\varepsilon^2} ,$$

where

$$H(X) = -E \log p(X)$$

and

$$\text{var}\left[\frac{-1}{n}\log p(X_1,\dots,X_n)\right] = \frac{n\sigma^2}{n^2} = \frac{\sigma^2}{n},$$

where

$$\sigma^2 = \sum_{x\in X} p(x)(\log p(x))^2 - \left(\sum_{x\in X} p(x)\log p(x)\right)^2$$

$$\therefore \quad P_r\left[\left|\frac{-1}{n}\log p(X_1,\dots,X_n) - H(X)\right| \geq \varepsilon\right] \leq \frac{\sigma^2}{n\varepsilon^2}$$

$$\therefore \quad \lim_{n\to\infty} P_r[\quad] = 0,$$

and the theorem follows. □

Thus, by Chebychev's Inequality, with high probability,

$$\left|\frac{-1}{n}\log p(x_1, x_2,\dots,x_n) - H(X)\right| \leq \varepsilon,$$

or removing the magnitude when the argument is positive,

$$p(x_1,\dots,x_n) \geq 2^{-n(H+\varepsilon)} \tag{3.26}$$

and removing the magnitude when the argument is negative,

$$p(x_1,\dots,x_n) \leq 2^{-n(H-\varepsilon)} \tag{3.27}$$

These two results imply that

$$2^{-n(H(X)+\epsilon)} \leq p(x_1, x_2,\dots,x_n) \leq 2^{-n(H(X)-\epsilon)} \tag{3.28}$$

and therefore that the probability assigned to an observed sequence should be about 2^{-nH}.

This observation is used to divide the set of all sequences into two sets, the typical set, where the sample entropy is close to the true entropy, and the non-typical set, which contains all of the other sequences. Further, any property that is true for the typical sequence will then be true with high probability and can be associated with the average behavior of a large sample. Notice, however, that the result as shown here does not imply that all sequences in a typical set have the "right" probability, but that the sample entropy is close to the true entropy.

A number of properties can be demonstrated directly from the AEP. In particular, in can be shown that, (1) the typical set has probability nearly 1, (2) all elements of the typical set are nearly equiprobable, and (3) the number of elements in the typical set is nearly 2^{nH}.

The question that naturally arises is "What does this have to do with data compression or lossless source coding?" The answer is as follows. We want to encode a set $X_1, X_2,\dots,X_n$ of i.i.d.

random variables with pmf $p(x)$. The approach is to separate all possible such sequences into two sets: the typical set $A_\epsilon^{(n)}$ and its complement $A_\epsilon^{(n)^c}$, the atypical set.

We order all elements in each set according to some order. Then we can represent each sequence of $A_\epsilon^{(n)}$ by giving the index of the corresponding element in the set. We need to assign a code to each sequence contained in both sets. There are $\leq 2^{n(H+\epsilon)}$ sequences in the typical set, $A_\epsilon^{(n)}$, so no more than $n(H + \epsilon) + 1$ bits are needed (the extra bit is there in case $n(H + \epsilon)$ is not an integer). We prefix all of these sequences by a 1, so that the total length for each sequence in the typical set is $\leq n(H + \epsilon) + 2$ bits .

We do not have an expression for the number of elements in the atypical set, but perhaps surprisingly, this does not matter because of its asymptotically vanishing probability. Therefore, we assign codes for the atypical set by counting all possible length n sequences, and assign codes of length $n \log |\mathcal{X}| + 1$ bits plus 1 since we add a prefix of a 0.

We now have a code for all possible sequences in X^n.

This somewhat strangely constructed code has the nice property that the codeword length of the typical sequences is about nH. Further, the first bit tells the decoder how many bits follow in a codeword, and the resulting code is uniquely decodable.

Using the properties of the typical set and the coding method just described, it can be shown that with $l(x^n)$ representing the length of the codeword for a sequence x^n,

$$E\left[\frac{1}{n}l(X^n)\right] \leq H(X) + \epsilon ,$$

for n sufficiently large.

Thus, we can represent sequences X^n using no more than $nH(X)$ bits on the average, and we have another proof of Shannon's theorem for lossless (noiseless) source coding.

3.8   NOTES AND ADDITIONAL REFERENCES

The coverage of this chapter combines the standard topics of the lossless source coding theorem, the Kraft inequality, the AEP, and Huffman coding, with Elias coding (and a hint at arithmetic coding), and the much less common topic of Lempel-Ziv coding methods. Since Lempel-Ziv techniques are universal coding methods, they are often not included in the same chapter as Huffman coding, for example, but are included here because of their importance in applications and the lack of a separate chapter on universal coding. Proofs of the Source Coding Theorem are available in Gallager [4], Cover and Thomas [5], Blahut [6], and McEliece [14], all of them a little different.

CHAPTER 4

Channel Capacity

4.1 THE DEFINITION OF CHANNEL CAPACITY

We now turn our attention to the problem of communicating source information over a noisy channel [4, 5]. With respect to the general communication system block diagram in Fig. 1.2, we are presently interested in the channel encoder, channel, and channel decoder blocks. For our current purposes, it is of no interest whether the source encoder/decoder blocks are present or whether the source output is connected directly to the channel encoder input and the channel decoder output is passed directly to the user. What we are interested in here is the transmission of information over a noisy channel. More specifically, we would like to address the question: Given the characterization of a communications channel (a model), what is the maximum bit rate that can be sent over this channel with negligibly small error probability? We find that the mutual information between the channel input and output random variables plays an important role in providing the answer to this question.

To introduce the approach and the fundamental ideas, we begin by considering *discrete memoryless channels* that have finite input and output alphabets and for which the output letter at any given time depends only on the channel input letter at the same time instant. Therefore, with reference to Fig. 1.2, we define a *discrete memoryless channel* (DMC) with input alphabet $W = \{1, 2, \ldots, M\}$ and probability assignment $P_W(w), w = 1, 2, \ldots, M$, and with output alphabet $X = \{1, 2, \ldots, N\}$ and *transition probabilities* $P_{X|W}(x|w), w = 1, 2, \ldots, M$, and $x = 1, 2, \ldots, N$. From these quantities we can calculate the mutual information between the channel input W and output X. Then, the Channel Capacity of a DMC with input W and output X is defined as

$$
\begin{aligned}
C &\triangleq \max_{\text{all } P_W(\cdot)} I(W; X) \\
&= \max_{\text{all } P_W(\cdot)} \sum_{w=1}^{M} \sum_{x=1}^{N} P_{WX}(w, x) \log \frac{P_{WX}(w, x)}{P_W(w) P_X(x)},
\end{aligned}
\tag{4.1}
$$

where the maximum is taken over all channel input probability assignments.

We note that $I(W; X)$ is a function of the input probabilities and the transition probabilities, whereas the channel capacity C is a function of the input probabilities only. Once we have selected a physically appropriate model for the channel, we assume that we have no control over the channel transition probabilities, since they are determined by the channel behavior as represented by the model. In words, Eq. (4.1) says that the capacity of a DMC is the largest mutual information between the input and output that can be transmitted over the channel in one use. The physical significance of the channel capacity expression is illustrated by the following theorem.

Theorem 4.1   Shannon's Second Theorem (Channel Coding Theorem). *[1, 4, 5] Capacity C expressed as*

$$C = \max_{p(x)} I(W; X) . \tag{4.2}$$

is the maximum rate that we can send information over the channel and recover the information at the channel output with vanishingly small error probability.

We present very few detailed proofs in this book, but later in the current chapter we provide an outline of a proof because of the physical insights that can be gained from the basic approach. It is emphasized again that the mathematical expression for capacity in terms of mutual information only has physical meaning because of this theorem. It is often difficult for students to understand that a mathematical proof can give a physically meaningful result, but such is the case here and elsewhere in information theory and rate distortion theory. The physical significance of the mathematical capacity expression that results from the maximization depends on the accuracy of the channel model in representing the actual physical channel.

This leads to the next critical point: Channel capacity is based on a chosen *model* for the channel, as represented here by the transition probabilities $P_{X|W}(x|w), w = 1, 2, \ldots, M$, and $x = 1, 2, \ldots, N$. If the model changes, the capacity result will change. This highlights the fact that in order for information theory (and rate distortion theory) to be physically significant, the engineer performing the analyses must also have a strong grasp of the physical problem of interest.

Therefore, for the chosen channel model, Channel Capacity is a fundamental limit on the maximum data rate that can be transmitted reliably over the channel. As a consequence, Channel Capacity is a physically significant quantity and has a myriad of uses. One such use is to allow a communications engineer to determine how close current or proposed approaches for communication over the channel are to the fundamental limit on the reliable transmitted rate. The principal implication being that if one is already operating close to channel capacity, perhaps additional effort or added complexity is not warranted.

To calculate the channel capacity, it is necessary, as indicated by Eq. (4.1), to perform a maximization over M variables, the $P_W(w)$, subject to the constraints that $P_W(w) \geq 0$ for all w and $\sum_{w=1}^{M} P_W(w) = 1$. In general, this is a difficult task. The following example illustrates the calculation of capacity for a very simple channel model, the *binary symmetric channel* (BSC), which can be performed by hand using calculus. Later in the chapter, we present other approaches for obtaining channel capacity expressions for other channel models.

Example 4.2   A special case of a DMC is the BSC with binary input and output alphabets $\{0, 1\}$ and transition probabilities $P_{X|W}(0|0) = P_{X|W}(1|1) = 1 - p$ and $P_{X|W}(0|1) = P_{X|W}(1|0) = p$, where W is the input random variable and X is the output random variable. A standard diagram for the BSC is shown in Fig. 4.1. We can rewrite the mutual information as

$$I(W; X) = \sum_{x=1}^{N} \sum_{w=1}^{M} P_{X|W}(x|w) P_W(w) \log \frac{P_{X|W}(x|w)}{\sum_{w=1}^{M} P_{X|W}(x|w) P_W(w)}. \qquad (4.3)$$

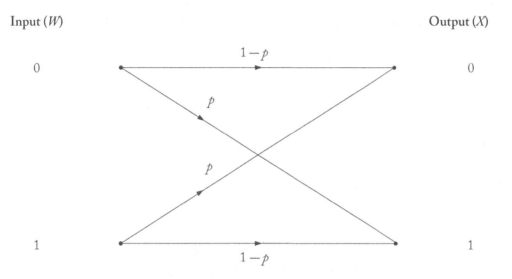

Input (W)   Output (X)

0    $1-p$ 0

p

p

1    $1-p$ 1

Figure 4.1: Binary symmetric channel (BSC) (see Ex. 4.1).

For the current example $M = N = 2$, $X = \{0, 1\}$, and $W = \{0, 1\}$. We need to write $I(W; X)$ in terms of $P_W(0)$ and $P_W(1) = 1 - P_W(0)$ so that we can perform the maximization. First, we evaluate the denominators of the argument of the logarithm as

$$x = 0 : \sum_{w=1}^{2} P_{X|W}(0|w) P_W(w) = (1 - p) P_W(0) + p P_W(1)$$

$$= (1 - 2p) P_W(0) + p \triangleq \mathrm{den}_0$$

$$x = 1 : \sum_{w=1}^{2} P_{X|W}(1|w) P_W(w) = p P_W(0) + (1 - p) P_W(1)$$

$$= 1 - p - (1 - 2p) P_W(0) \triangleq \mathrm{den}_1,$$

since $P_W(1) = 1 - P_W(0)$. Now, expanding Eq. (4.3), we have

$$
\begin{aligned}
I(W;X) &= (1-p)P_W(0)\log\frac{1-p}{\text{den}_0} + pP_W(1)\log\frac{p}{\text{den}_0} \\
&\quad + pP_W(0)\log\frac{p}{\text{den}_1} + (1-p)P_W(1)\log\frac{1-p}{\text{den}_1} \\
&= p\log p + (1-p)\log(1-p) + \big[2pP_W(0) - P_W(0) - p\big]\log(\text{den}_0) \\
&\quad + \big[p + P_W(0) - 2pP_W(0) - 1\big]\log(\text{den}_1),
\end{aligned} \tag{4.4}
$$

where the last equality results from letting $P_W(1) = 1 - P_W(0)$ and simplifying.

Taking the partial derivative of $I(W;X)$ with respect to $P_W(0)$ yields

$$
\begin{aligned}
\frac{\partial}{\partial P_W(0)} I(W;X) &= (2p-1)\log\text{den}_0 + \frac{2pP_W(0) - P_W(0) - p}{\text{den}_0}(1-2p) \\
&\quad + (1-2p)\log\text{den}_1 + \frac{p + P_W(0) - 2pP_W(0) - 1}{\text{den}_1}(2p-1) \\
&= -(1-2p)\log\text{den}_0 + (1-2p)\log\text{den}_1,
\end{aligned} \tag{4.5}
$$

where the last simplification follows after using the definitions of den_0 and den_1. Equating the partial derivative to 0, we find that $P_W(0) = \frac{1}{2}$. Upon substituting this value back into Eq. (4.4), the capacity of the BSC is found to be

$$
C = 1 + p\log p + (1-p)\log(1-p) \tag{4.6}
$$

and is achieved with equally likely inputs. Thus, the conclusion is that for the BSC, the maximum rate in bits/letter that can be sent over the channel is achieved if we design the communications system such that the input to the channel consists of equally likely binary digits. Note that the input probabilities to the channel are quantities over which we have control by our communication system designs, in this case error control coding methods. Further, the capacity can be explicitly calculated once we specify the transition probability p of the channel.

Although, we have not shown that the partial derivative yields a maximum as opposed to a minimum, $I(W;X)$ is a concave or convex $\cap$ (read "cap") function of $P_W(\cdot)$, and hence we have found a maximum. This and other important properties of $I(W;X)$ are discussed in a later section and proofs are available elsewhere [14].

4.2 PROPERTIES OF CHANNEL CAPACITY

There are several key properties of mutual information and channel capacity that are helpful in evaluating channel capacity for a chosen channel model or for performing analyses that reveal fundamental insights for communication systems [4, 5]. A few of these are presented now:

1. $C \geq 0$, since $I(X;Y) \geq 0$.

2. $C \leq \log |\mathcal{X}|$ since $C = \max\limits_{p(x)} I(X;Y) \leq \max H(X) = \log |\mathcal{X}|$.

3. $C \leq \log |\mathcal{Y}|$.

4. $I(X;Y)$ is a continuous function of $p(x)$.

5. $I(X;Y)$ is a concave function of $p(x)$.

Since $I(X;Y)$ is a concave function over a closed convex set, a local maximum is a global maximum. From Properties 2 and 3, the maximum is finite and we can use maximum rather than supremum (sup).

The following example illustrates Property 3.

Example 4.3   For the BSC in Example 4.2 plot $I(W;X)$ as a function of p when $P_W(0) = \frac{3}{4}$ and $P_W(1) = \frac{1}{4}$ and compare to a plot of C given by Eq. (4.6). >From Eq. (4.4),

$$
\begin{aligned}
I(W;X) = {}& p \log p + (1-p) \log(1-p) \\
& + [2pP_W(0) - P_W(0) - p] \log[(1-2p)P_W(0) + p] \\
& + [p + P_W(0) - 2pP_W(0) - 1] \log[1 - p - (1-2p)P_W(0)] .
\end{aligned}
$$

Letting $P_W(0) = 3/4$,

$$
\begin{aligned}
I(W;X) = {}& p \log p + (1-p) \log(1-p) + \left[\frac{3}{2}p - \frac{3}{4} - p\right] \log\left[(1-2p)\left(\frac{3}{4}\right) + p\right] \\
& + \left[p + \frac{3}{4} - \frac{3}{2}p - 1\right] \log\left[1 - p - (1-2p)\left(\frac{3}{4}\right)\right] \\
= {}& p \log p + (1-p) \log(1-p) + \left[\frac{p}{2} - \frac{3}{4}\right] \log\left[\frac{3}{4} - \frac{p}{2}\right] \\
& + \left[-\frac{p}{2} - \frac{1}{4}\right] \log\left[\frac{1}{4} + \frac{p}{2}\right] .
\end{aligned}
$$

This is plotted on the following figure. Note that it is still symmetric about $p = 0.5$, but that the maximum is 0.81, whereas in Ex. 4.2 the maximum is 1.0. Not only does this example illustrate Property 3, it also shows that capacity is reduced when the input probabilities differ from the optimizing equally likely assignment.

4.3   CALCULATING CAPACITY FOR DISCRETE MEMORYLESS CHANNELS

It is often difficult to know how to begin to find capacity once a meaningful channel model has been chosen [4, 5]. One way to simplify the calculation of channel capacity is to exploit structure in the channel model. An important example of such structure is symmetry in the channel

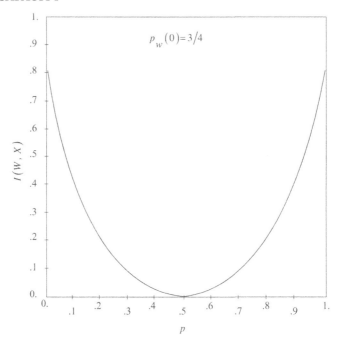

Figure 4.2: Mutual information for a BSC with unequal probability inputs.

probability transition matrix (PTM). The PTM is defined as the matrix of channel transition probabilities where the inputs are rows and the outputs are columns. A *discrete channel* is described by an input alphabet $\mathcal{X}$, an output alphabet $\mathcal{Y}$, and a probability transition matrix $[p(y|x)]$ specifying the input-output relationship. A discrete channel is said to be *memoryless* if the probability distribution of the output depends only upon the input at that time and is conditionally independent of previous channel inputs or outputs.

This last definition refers to a discrete memoryless channel (DMC) used without feedback. When we say DMC, we shall mean (unless explicitly stated otherwise) a DMC used without feedback. Letting $X^n = (x_1, x_2, \ldots, x_n)$ be an input sequence and $Y^n = (y_1, y_2, \ldots, y_n)$ be an output sequence, the nth use of a DMC (used without feedback) is described by

$$p(Y^n|X^n) = \prod_{i=1}^{n} p(y_i|x_i) . \tag{4.7}$$

Some Examples of DMCs and their corresponding PTM are:

Binary Symmetric Channel (BSC) PTM

$$\begin{bmatrix} 1-\alpha & \alpha \\ \alpha & 1-\alpha \end{bmatrix} .$$

The BSC is a very simple channel model that is surprisingly useful since it involves binary inputs and bit-flipping with a given probability.

Binary Erasure Channel (BEC) PTM

$$\begin{bmatrix} 1-\alpha & \alpha & 0 \\ 0 & \alpha & 1-\alpha \end{bmatrix}.$$

The BEC can represent a physical situation where the input consists of a binary digit but the bit is either deleted or the receiver cannot decide which bit was transmitted, so an erasure is assigned rather than decide the other bit was transmitted.

Ternary Symmetric Channel PTM

$$\begin{bmatrix} 1-\alpha-\beta & \alpha & \beta \\ \alpha & 1-\alpha-\beta & \beta \\ \beta & \alpha & 1-\alpha-\beta \end{bmatrix}.$$

The ternary symmetric channel can represent the physical situation where, for example, transmitted pulses can take on three possible values, $+V$, 0, and $-V$, and the values are detected by the receiver with the probabilities shown.

Two of these channel labels involve the word "symmetric", and so it is noted that definitions of symmetry differ across the several textbooks in the field, and the reader is cautioned to look carefully at the definitions here and elsewhere before using the accompanying results. We define symmetric, weakly symmetric, and partitioned symmetric channels [4, 5, 14].

Definition 4.4   A channel is said to be *symmetric* if the rows of the channel PTM are permutations of each other, and the columns are permutations of each other.

Definition 4.5   A channel is said to be *weakly symmetric* if every row of the PTM $p(\cdot|x)$ is a permutation of every other row, and all the column sums $\sum_{x} p(y|x)$ are equal.

Definition 4.6   A DMC is defined to be *partitioned symmetric* if the set of outputs can be partitioned into subsets in such a way that for each subset, the matrix of transition probabilities (using inputs as rows and outputs of the subsets as columns) has the property that each row is a permutation of each other row and each column is a permutation of each other column.

Note that the BSC satisfies all three definitions of symmetry, the BEC is seen to be partitioned symmetric, and the ternary symmetric channel is weakly symmetric.

For symmetric and weakly symmetric channels, $\sum_y p(y|x) \log p(y|x)$ is independent of x, so we have

$$
\begin{aligned}
I(X;Y) &= H(Y) - H(Y|X) \\
&= H(Y) + \sum_x p(x) \sum_y p(y|x) \log p(y|x) \\
&= H(Y) - H(r) \\
&\leq \log |\mathcal{Y}| - H(r) \,,
\end{aligned}
\tag{4.8}
$$

where r is a row of the PTM. Since $H(r)$ is independent of x, the maximum is achieved when the outputs are equally likely.

Equality is achieved for both symmetric and weakly symmetric channels if $p(x) = \frac{1}{|\mathcal{X}|}$ since

$$
\begin{aligned}
p(y) &= \sum_x p(y|x) p(x) = \frac{1}{|\mathcal{X}|} \sum_x p(y|x) \\
&= \frac{c}{|\mathcal{X}|} \quad \text{for all } y \in Y \,.
\end{aligned}
$$

Theorem 4.7   *For a partitioned symmetric DMC, capacity is achieved by using the inputs with equal probability.*

The following proof can be skipped without loss of continuity.

Proof. Consider

$$
I(x = k; Y) = \sum_y p(y|x) \log \frac{p(y|x)}{\frac{1}{|\mathcal{X}|} \sum_\alpha p(y|\alpha)} \,.
$$

Within a partition, $\frac{1}{|\mathcal{X}|} \sum_\alpha p(y|\alpha)$ is the same for all y since the sum is over columns and they are permutations of each other.

Thus, $I(x = k; y)$ has the same $I(x = k; y) = \log[p(y|k)/\frac{1}{|\mathcal{X}|} \sum_x p(y|x)]$ permutation properties as $p(y|k)$, and hence so does

$$
p(y|k) I(x = k; y) = \beta_{ky} \,.
$$

Consider,

$$I(x = k; y) = \sum_y p(y|k) I(x = k; y)$$

$$= \sum_y p(y|k) \log \frac{p(y|k)}{\frac{1}{|X|} \sum_x p(y|x)}$$

$$= \sum_y \beta_{ky} = \underbrace{\sum_{y_1} \beta_{ky}}_{} + \underbrace{\sum_{y_2} \beta_{ky}}_{} + \cdots + \underbrace{\sum_{y_n} \beta_{ky}}_{}$$

$$\underbrace{\phantom{\sum_{y_1} \beta_{ky} + \sum_{y_2} \beta_{ky} + \cdots + \sum_{y_n} \beta_{ky}}}_{\text{partitions}}$$

$$\boxed{\text{We use the partitioning here}} = \underbrace{C_1 + C_2 + \cdots + C_n}_{\text{sums over rows}} = \underbrace{C \text{ for all } k}_{\text{since rows are permutations}}$$

Thus, the earlier theorem is satisfied and capacity is achieved.   □

Considering the ternary symmetric channel PTM, we see that the rows of the whole matrix are permutations, but no partitioning works.

An often useful approach to finding channel capacity for discrete memoryless channels relies on the following theorem. The solution is one example of applying the Kuhn-Tucker conditions.

Theorem 4.8   *A DMC has input W and output X with transition probabilities $P_{X|W}(x|w)$, $j = 1, \ldots, M$, $x = 1, 2, \ldots, N$. Necessary and sufficient conditions for a set of input probabilities $P_W(w)$, $w = 1, 2, \ldots, M$, to achieve capacity is for*

$$I(W = w; X) = C \qquad \text{for all } w \text{ with } P_W(w) > 0$$

and

$$I(W = w; X) \leq C \qquad \text{for all } w \text{ with } P_W(w) = 0,$$

for some number C, where

$$I(W = w; X) = \sum_{x=1}^{N} P_{X|W}(x|w) \log \frac{P_{X|W}(x|w)}{\sum_{w=1}^{M} P_W(w) P_{X|W}(x|w)}.$$

The number C is the channel capacity.

The main use of this theorem is to check the validity of some hypothesized set of input probabilities; that is, guess and verify. Thus, for the BSC in Ex. 4.2, we might guess by symmetry that $P_W(0) = P_W(1) = \frac{1}{2}$ achieves capacity.

Example 4.9   Substantiate the claim that $P_W(0) = P_W(1) = \frac{1}{2}$ achieves capacity for the BSC and find the capacity in Eq. (4.6) by using the immediately preceding theorem. Guess $P_W(0) =$

$P_W(1) = 1/2$ and compute $I(w = 0; X)$ and $I(w = 1; X)$. Thus,

$$
\begin{aligned}
I(w = 0; X) &= \sum_{x=1}^{2} P_{X|W}(x|0) \log \frac{P_{X|W}(x|0)}{\frac{1}{2}[P(x|0) + P(x|1)]} \\
&= P_{X|W}(0|0) \log 2 P_{X|W}(0|0) + P_{X|W}(1|0) \log 2 P_{X|W}(1|0) \\
&= (1 - p) \log 2(1 - p) + p \log 2p \ ;
\end{aligned}
$$

and

$$
\begin{aligned}
I(w = 1; X) &= \sum_{x=1}^{2} P_{X|W}(x|1) \log \frac{P_{X|W}(x|1)}{\frac{1}{2}[P(x|0) + P(x|1)]} \\
&= P_{X|W}(0|1) \log 2 P_{X|W}(0|1) + P_{X|W}(1|1) \log 2 P_{X|W}(1|1) \\
&= p \log 2p + (1 - p) \log 2(1 - p) \ .
\end{aligned}
$$

Thus,

$$
I(w = 0; X) = I(w = 1; X)
$$

which by the theorem must be C.

Hence,

$$
\begin{aligned}
C &= p \log 2p + (1 - p) \log 2(1 - p) \\
&= p \log 2 + p \log p + (1 - p) \log 2 + (1 - p) \log(1 - p) \ .
\end{aligned}
$$

For base 2 logs,

$$
\begin{aligned}
C &= p + p \log p + 1 - p + (1 - p) \log(1 - p) \\
&= 1 + p \log p + (1 - p) \log(1 - p)
\end{aligned}
$$

which is the same as Eq. (4.6).

Example 4.10   Use the preceding theorem to show that the capacity of the binary erasure channel is α. *Hint:* Guess equally likely inputs and verify.

For $P_W(1) = P_W(2) = 1/2$, then

$$
\begin{aligned}
I(W = w; X) &= \sum_{x=1}^{3} P_{X|W}(x|w) \log \left\{ \frac{P_{X|W}(x|w)}{\sum_{w=1}^{2} P_W(w) P_{X|W}(x|w)} \right\} \\
&= P_{X|W}(2|1) \log \frac{P_{X|W}(2|1)}{P_X(2)} + P_{X|W}(1|1) \log \frac{P_{X|W}(1|1)}{P_X(1)} \\
&\quad + P_{X|W}(3|1) \log \frac{P_{X|W}(3|1)}{P_X(3)} \\
&= 0 \log 0 + \alpha \log \frac{\alpha}{P_X(1)} + (1 - \alpha) \log \frac{1 - \alpha}{1 - \alpha} \ .
\end{aligned}
$$

Now, $P_X(1) = \frac{\alpha}{2} = P_X(2)$, so

$$I(w = 1; X) = \alpha \log \frac{\alpha}{(\alpha/2)} = \alpha \ .$$

Further,

$$I(w = 2; X) = \alpha \ ,$$

so by the preceding theorem,

$$C = \alpha \ .$$

We now have a couple of approaches for obtaining expressions for channel capacity and we have given some examples for simple DMCs. The Kuhn-Tucker conditions are presented in the appendices and are used in several places throughout the book since we are interested in constrained optimization problems; that is, our goal is to find the best performance theoretically achievable. A note for the reader: In the information theory and rate distortion theory literature, authors often use the substitution OPTA to stand for Optimum Performance Theoretically Achievable. We do not use this terminology elsewhere in the book.

Let a length n sequence, denoted as X^n, be the input to a DMC, and let Y^n be the corresponding output sequence. Then

$$I(X^n; Y^n) \leq nC \tag{4.9}$$

for all $p(x^n)$.

Proof.

$$I(X^n; Y^n) = H(Y^n) - H(Y^n|X^n) \tag{4.10}$$

$$= H(Y^n) - \sum_{i=1}^{n} H(Y_i|Y_1, \ldots, Y_{i-1}, X^n) \tag{4.11}$$

$$= H(Y^n) - \sum_{i=1}^{n} H(Y_i|X_i) \tag{4.12}$$

since for a DMC, each output depends only on the corresponding input. We continue by noting that the entropy of the output sequence is less than the sum of the individual entropies,

$$\leq \sum_{i=1}^{n} H(Y_i) - \sum_{i=1}^{n} H(Y_i|X_i)$$

$$= \sum_{i=1}^{n} I(X_i; Y_i) \leq nC \ . \tag{4.13}$$

since capacity is the maximum of the mutual information. □

Properties of this type are often useful in getting some quick insights into channel capacity for various channels even when the explicit calculation of channel capacity may be difficult.

4.4   THE CHANNEL CODING THEOREM

Shannon introduced the idea of channel capacity in 1948, when he defined capacity as the maximum rate that one can transmit over a channel with an arbitrarily small error probability [5]. Conceptually, this would seem impossible; that is, to achieve an arbitrarily small error probability for a noisy channel. However, Shannon based his development of capacity on some extraordinarily innovative new insights.

First, he was trying to understand what is possible, separate from how to do it. Specifically, he was after an existence proof; he wanted to show that there *was* a codebook that achieved an arbitrarily small error probability. Second, he was willing to trade delay for performance. Today, we would equate increased delay with increased signal processing complexity. Third, Shannon considered random codebooks, including all possible good and bad codebooks, and fourth, he allowed asymptotically long codewords. He then considered the probability of error averaged over all possible codebooks, good and bad. This approach combined with the arbitrarily long codewords allowed him to show that the average error probability over all codebooks has an exponentially decreasing error probability. Finally, he observed that if this last fact were true, then there must be at least one good codebook in the set. This collection of ideas was nothing short of revolutionary. Today, we might call Shannon's proof of channel capacity a disruptive technology, and that would be an understatement.

The concept of channel capacity and the calculation of channel capacity for specific channel models, is often criticized for being "only" a proof of the existence of a good codebook. Of course, deriving an expression for capacity does, in fact, only prove the existence of a good code and does not explicitly exhibit a good codebook design. However, knowing the best that can be done and comparing the best possible performance with the best performance available with current designs is a great deal of useful knowledge in and of itself. Second, proofs of channel capacity can often provide insights into the structure (or lack thereof) of good codes. When appropriate, these are pointed out in the following.

Given the list of Shannon's insights and the basic ideas behind channel capacity, it is likely still not evident to the reader how to go about proving the channel coding theorem nor why the mutual information turns out to be an important quantity. In the following, we first outline the required proof and align each step with Shannon's assumptions, and then we explain how mutual information comes to play such an important role in the expression for capacity.

Theorem 4.11   Shannon's Second Theorem (Channel Coding Theorem).   *[1, 4, 5] Capacity C expressed as*

$$C = \max_{p(x)} I(X;Y) . \tag{4.14}$$

is the maximum rate that we can send information over the channel and recover the information at the channel output with vanishingly small error probability. More specifically, if the transmission rate $R < C$, then there exists a sequence of length-n codes with error probability $P_r^{(n)}(\varepsilon)$ that approaches

0 *as n increases, and conversely, for any sequence of codes with* $P_r^{(n)}(\mathcal{E}) \to 0$ *with increasing n, then* $R \le C$.

Outline of the Proof that rates R < C are achievable: If we have M messages to communicate length n sequences, then the rate R of the code is related to M by $M = 2^{nR}$, and thus there are 2^{nR} codewords.

For a fixed $p(x)$, we independently generate length n codewords according to the distribution

$$p(x^n) = \prod_{i=1}^{n} p(x_i) \, . \tag{4.15}$$

so that the probability of a particular code $\mathcal{C}$ is

$$P_r(\mathcal{C}) = \prod_{w=1}^{2^{nR}} \prod_{i=1}^{n} p(x_i(w)) \, . \tag{4.16}$$

where each element is generated iid according to $p(x)$.

With these steps as setup, the communication system operates as follows:

A random code $\mathcal{C}$ is designed (generated) by (4.16) according to $p(x)$, and the code is known to both the transmitter and receiver. The transmitter and receiver are also assumed to know the channel transition matrix $p(y|x)$.

A message W to be sent is chosen according to a uniform distribution,

$$P_r(W = w) = 2^{-nR}, \qquad w = 1, 2, \ldots, 2^{nR} \, . \tag{4.17}$$

and the wth codeword $X^n(w)$ is sent over the channel.

A sequence Y^n arrives at the receiver according to the distribution

$$P(y^n|x^n(w)) = \prod_{i=1}^{n} p(y_i|x_i(w)) \, . \tag{4.18}$$

The receiver decodes the received sequence as $\hat{W}$, an estimate of the input message W. There is a decoding error if $\hat{W} \ne W$.

Let $\mathcal{E}$ be the event $\{\hat{W} \ne W\}$.

We can express the average probability of error, averaged over all codewords in the codebook, and averaged over all codebooks, as

$$P_r(\mathcal{E}) = \sum_{\mathcal{C}} P(\mathcal{C}) P_e^{(n)}(\mathcal{C}) \tag{4.19}$$

$$= \sum_{\mathcal{C}} P(\mathcal{C}) \frac{1}{2^{nR}} \sum_{w=1}^{2^{nR}} p_w(\mathcal{C}) \tag{4.20}$$

$$= \frac{1}{2^{nR}} \sum_{w=1}^{2^{nR}} \sum_{\mathcal{C}} P(\mathcal{C}) p_w(\mathcal{C}) , \tag{4.21}$$

where $P_e^{(n)}(\mathcal{C})$ is defined for whatever decoding rule is implemented (we will choose a decoding rule and elaborate on this statement shortly). It is this decoding error that we must prove is exponentially small with increasing n.

So, the next question that we must address is what is the form of the decoder? The most common proofs of the channel coding theorem use either maximum likelihood decoding or decoding using jointly typical sequences. Maximum likelihood (ML) decoding is a structure that is very familiar to communications systems designers, so it would appear that ML decoding would be preferable. However, it turns out that the analysis using ML decoding is more difficult mathematically and nonintuitive.

Decoding using jointly typical sequences is likely a new concept and decoding using joint typicality is more difficult to envision as a physical receiver structure, but the proof of the channel coding theorem is more straightforward and produces a more intuitive result than the ML decoder approach. Furthermore, although decoding using joint typicality is suboptimal, it allows us to show that the average error probability asymptotically approaches zero and that is all that we need.

In the next section, we introduce the definition of jointly typical sequences and the joint Asymptotic Equipartition Property (AEP), and show how these can lead to the desired result.

4.5   DECODING AND JOINTLY TYPICAL SEQUENCES

Given a channel output Y^n, we need to decide what input sequence was transmitted [5]. Each input sequence is represented for transmission over the channel by a codeword, so if the message to be transmitted has index i, then the transmitted codeword is $X^n(i)$. Using joint typicality, we decode a channel output Y^n as the ith index if the codeword $X^n(i)$ is "jointly typical" with the received signal Y^n. To analyze the error probability, we define the concept of joint typicality and find the probability of joint typicality when Y^n is due to the input codeword $X^n(i)$ and the probability of Y^n occurring when $X^n(i)$ is not the transmitted codeword.

Definition 4.12   The set $A_\epsilon^{(n)}$ of jointly typical sequences $\{(x^n, y^n)\}$ with respect to the distribution $p(x, y)$ is the set of n-sequences with empirical entropies within ϵ of the true entropies, i.e.,

$$A_\epsilon^{(n)} = \Big\{ (x^n, y^n) \epsilon X^n \times Y^n : \tag{4.22}$$

$$\Big| \frac{-1}{n} \log p(x^n) - H(X) \Big| < \epsilon, \tag{4.23}$$

$$\Big| \frac{-1}{n} \log p(y^n) - H(Y) \Big| < \epsilon, \tag{4.24}$$

$$\Big| \frac{-1}{n} \log p(x^n, y^n) - H(X, Y) \Big| < \epsilon \Big\}, \tag{4.25}$$

where

$$p(x^n, y^n) = \prod_{i=1}^{n} p(x_i; y_i). \tag{4.26}$$

Given this definition, the following results can be obtained.

Theorem 4.13   (Joint AEP).   *[5] Let (X^n, Y^n) be sequences of length n drawn i.i.d. according to $p(x^n, y^n) = \prod_{i=1}^{n} p(x_i, y_i)$. Then*

1. *$\Pr((X^n, Y^n) \epsilon A_\epsilon^{(n)}) \to 1$ as $n \to \infty$.*

2. *$|A_\epsilon^{(n)}| \le 2^{n(H(X,Y)+\epsilon)}$.*

3. *If $(\tilde{X}^n, \tilde{Y}^n) \sim p(x^n)p(y^n)$, i.e., $\tilde{X}^n$, and $\tilde{Y}^n$ are independent with the same marginals as $p(x^n, y^n)$, then*

$$\Pr((\tilde{X}^n, \tilde{Y}^n) \epsilon A_\epsilon^{(n)}) \le 2^{-n(I(X;Y)-3\epsilon)}. \tag{4.27}$$

Also, for sufficiently large n,

$$\Pr((\tilde{X}^n, \tilde{Y}^n) \epsilon A_\epsilon^{(n)}) \ge (1 - \epsilon)2^{-n(I(X;Y)+3\epsilon)}.$$

The proofs are left to the references. However, we can now provide some motivation for the role of mutual information in the calculation of channel capacity.

From the joint AEP, we see that the probability of the two statistically independent sequences (X^n, Y^n), with marginal distributions the same as the transmitted and received codewords, being jointly typical at the receiver is about $2^{-n(I(X;Y))}$ and so we see the appearance of the mutual information in conjunction with jointly typical decoding. Without going into further mathematical details concerning joint typicality, we can motivate channel capacity as follows.

Therefore, the total number of possible (typical) Y sequences is $\cong 2^{nH(Y)}$. For each (typical) input n-sequence, there are approximately $2^{nH(Y|X)}$ possible Y sequences at the receiver, all of them equally likely. For good decoding, we need to operate such that no two X sequences produce the same Y output sequence, or we cannot decode correctly.

Thus, the set of $2^{nH(Y)}$ sequences is to be divided into sets of size $2^{nH(Y|X)}$ corresponding to the different input sequences. Dividing, we see that the total number of disjoint sets is less than or equal to

$$2^{n(H(Y)-H(Y|X))} = 2^{nI(X;Y)} .$$

Hence, we can send at most $\cong 2^{nI(X;Y)}$ distinguishable sequences of length n, (or $\log 2^{nI(X;Y)} = nI(X;Y)$ bits), which is $I(X;Y)$ bits per channel use. Maximizing this quantity over $p(x)$ yields the capacity C.

4.6   FANO'S INEQUALITY AND THE CONVERSE TO THE CODING THEOREM

The achievability proof for the Channel Coding Theorem shows that if $R < C$, then the error probability can be made asymptotically small [4, 5, 6]. Now, we consider the converse which is if $R > C$ the error probability is bounded away from 0 for n large. The proof relies on what is called Fano's Inequality and we begin by stating and proving it for $n = 1$.

For $n = 1$, we have a message W which we encode as $X(W)$. At the decoder, we observe Y and decode the transmitted message as $\hat{W} = g(Y)$. Defining the probability of error as

$$P_e = \Pr(\hat{W} \neq W) . \tag{4.28}$$

and denoting the error event by the random variable $\mathcal{E}$, we define

$$\mathcal{E} = \begin{cases} 1, & \text{if } \hat{W} \neq W \\ 0, & \text{if } \hat{W} = W \end{cases} . \tag{4.29}$$

and consider $H(\mathcal{E}, W|Y)$, which we expand using the chain rule for entropies as

$$H(\mathcal{E}, W|Y) = H(W|Y) + H(\mathcal{E}|W, Y) \tag{4.30}$$
$$= H(\mathcal{E}|Y) + H(W|\mathcal{E}, Y) . \tag{4.31}$$

Working with the two versions on the right hand side, we immediately can simplify to

$$H(W|Y) \leq H(\mathcal{E}) + H(W|\mathcal{E}, Y) . \tag{4.32}$$

since $H(\mathcal{E}|W, Y) = 0$ because $\mathcal{E}$ is a function of W and $g(Y)$, and $H(\mathcal{E}|Y) \leq H(\mathcal{E})$.

The last term on the right can be expanded as

$$H(W|\mathcal{E}, Y) = P(\mathcal{E} = 0)H(W|Y, \mathcal{E} = 0) + P(\mathcal{E} = 1)H(W|Y, \mathcal{E} = 1) . \tag{4.33}$$

We note that since $\hat{W} = g(Y)$, if $\mathcal{E} = 0$ and Y is given, $\hat{W} = g(Y^n) = W$, and $H(W|Y, \mathcal{E} = 0) = 0$. Further, given $\mathcal{E} = 1$, $H(W|Y, \mathcal{E} = 1) \leq \log(|W| - 1)$.

Therefore, letting $P(\mathcal{E} = 1) = P_e$, we obtain

$$H(W|Y) \leq H(\mathcal{E}) + P_e \log(|W| - 1) . \tag{4.34}$$

This is the most familiar form of Fano's Inequality. Gallager points out a nice intuitive interpretation of this inequality [4]. The conditional uncertainty of W given Y is represented as two terms. One term is the uncertainty as to whether there has been an error, upper bounded by $H(\mathcal{E})$, and another term expresses the uncertainty about which message was sent if an error is made, upper bounded by $P_e \log(|W| - 1)$.

We can simplify further by noting that $H(\mathcal{E}) \leq 1$, so we get the weaker version

$$H(W|Y) \leq 1 + P_e \log(|W| - 1) . \tag{4.35}$$

We can now easily consider our more general problem.

Returning to the channel coding theorem setup, we have a codeword length n, we have a message W which we encode as $X^n(W)$. At the decoder, we observe Y^n and decode the transmitted message as $\hat{W} = g(Y^n)$. Defining the probability of error as

$$P_e = \Pr(\hat{W} \neq W) . \tag{4.36}$$

and denoting the error event by the random variable $\mathcal{E}$, we define the probability of error

$$P_e^{(n)} = \Pr(\hat{W} \neq W) . \tag{4.37}$$

and the error random variable as before. We can mimic the steps used in the special case of $n = 1$ to obtain

$$H(W|Y^n) \leq 1 + P_e^{(n)} \log(|W| - 1) \tag{4.38}$$

Working with the last term on the right, we write

$$\begin{aligned} P_e^{(n)} \log(|W| - 1) &\leq P_e^{(n)} \log(|W|) \\ &= P_e^{(n)} \log 2^{nR} \\ &= P_e^{(n)} nR . \end{aligned} \tag{4.39}$$

and combining the results produces

$$H(W|Y^n) \leq 1 + P_e^{(n)} nR . \tag{4.40}$$

Noting that $X^n(W)$ is a function of W,

$$H(X^n(W)|Y^n) \le H(W|Y^n) ,$$

and we have the final form of Fano's Inequality

$$H(X^n|Y^n) \le 1 + P_e^{(n)}nR . \tag{4.41}$$

We now prove the converse to the channel coding theorem, that is, any sequence of $(2^{nR}, n)$ codes with maximal error probability $\lambda^{(n)} \to 0$ must have $R \le C$.

Proof. If the maximal probability of error tends to zero, then the average probability of error for the sequence of codes also goes to zero, i.e., $\lambda^{(n)} \to 0$ implies $P_e^{(n)} \to 0$, for each n. Let W be uniformly distributed over $\{1, 2, \ldots, 2^{nR}\}$. Since W has a uniform distribution, $P_e^{(n)} = \Pr(\hat{W} \ne W)$, otherwise each value of W would be weighted differently. Hence,

$$nR = H(W) = H(W|Y^n) + I(W; Y^n) \tag{4.42}$$
$$\le H(W|Y^n) + I(X^n(W); Y^n) \tag{4.43}$$
$$\tag{4.44}$$

Using Fano's Inequality

$$nR \le 1 + P_e^{(n)}nR + I(X^n(W); Y^n) \tag{4.45}$$
$$\le 1 + P_e^{(n)}nR + nC \tag{4.46}$$

since capacity is the maximum of the mutual information and the successive inputs and outputs are independent. Dividing by n,

$$R \le \frac{1}{n} + P_e^{(n)}R + C . \tag{4.47}$$

Letting $n \to \infty$, the first two terms on the right $\to 0$ and

$$R \le C . \tag{4.48}$$

$\square$

We can also rewrite (4.47) as

$$P_e^{(n)} \ge 1 - \frac{1}{nR} - \frac{C}{R} . \tag{4.49}$$

This shows that if $R > C$, the probability of error is bounded away from 0 for large n. It follows that this conclusion holds for all n, since if $P_e^{(n)} = 0$ for small n, we can construct codes for large n with $P_e^{(n)} = 0$ by concatenating these codes.

This result that the average probability of error is bounded away from 0 for $R > C$., is called the *weak converse*. The *strong converse* says that at rates above capacity, the probability of error goes exponentially to 1. We do not treat the strong converse in this book.

4.7   THE ADDITIVE GAUSSIAN NOISE CHANNEL AND CAPACITY

An important model that is a building block for many important communications problems is the additive Gaussian noise channel given by

$$Y_i = X_i + Z_i \tag{4.50}$$

where X_i is the transmitted symbol and the noise $Z_i \sim \eta(0, N)$ is statistically independent of X_i [5, 29]. The input symbol X_i is often subject to an average (or peak) energy or power constraint, which fits many physical situations well. For any codeword $(x_1, x_2, \ldots, x_n)$ to be transmitted over the channel, the average power constraint can be expressed as

$$\frac{1}{n} \sum_{i=1}^{n} x_i^2 \leq P . \tag{4.51}$$

We can define the *information* capacity of the additive Gaussian noise channel with average power constraint P as

$$C = \max_{p(x):EX^2 \leq P} I(X;Y) . \tag{4.52}$$

and the capacity can be (and will be) shown to have the form

$$C = \frac{1}{2} \log \left(1 + \frac{P}{N} \right) . \tag{4.53}$$

Before deriving this expression and outlining the proof of achievability, we give a signal space type of justification based on a sphere packing argument.

Plausibility argument: Why we are able the construct $(2^{nC}, n)$ codes with low error probability:

Consider any codeword of length n. With high probability, the received vector is contained in a sphere of radius $\sqrt{n(N + \epsilon)}$ around the true codeword. If all received vectors in this sphere are assigned to the given codeword, there will be an error when this codeword is sent only if the received vector falls outside this sphere, which is an event with low probability. We choose the decoding spheres of other codewords similarly.

Those readers familiar with signal space, signal constellations, and error probability calculations in digital communications should note that the preceding and following development is very similar.

How many such codewords can we choose? The volume of an n-dimensional sphere is of the form $A_n r^n$ where r is the radius of the sphere.

In our case, each of the decoding spheres has radius $\sqrt{nN}$. The received vectors lie in a sphere of radius $\sqrt{n(P + N)}$. Thus, the maximum number of non-intersecting decoding spheres in this volume is no more than

$$\frac{A_n\left(\sqrt{n(P+N)}\right)^n}{A_n\left(\sqrt{nN}\right)^n} = \left(1+\frac{P}{N}\right)^{\frac{n}{2}}$$

$$= 2^{\frac{n}{2}\log(1+\frac{P}{N})} \tag{4.54}$$

so the code rate is $\frac{1}{2}\log(1+\frac{P}{N})$ bits/channel use. This is called a sphere packing argument.

To develop the capacity result from first principles, we start with the definition of mutual information and simplify as follows, We can calculate the information capacity as follows:

$$I(X;Y) = h(Y) - h(Y|X)$$
$$= h(Y) - h(X+Z|X)$$
$$= h(Y) - h(Z|X)$$
$$= h(Y) - h(Z)$$

since X and Z are statistically independent. Z is Gaussian so $h(Z) = \frac{1}{2}\log 2\pi e N$, and since Y is the sum of two independent random variables, $EY^2 = P + N$, using the average power constraint on X. We know that for a given variance, the Gaussian distribution maximizes entropy, so $h(Y) \le \frac{1}{2}\log 2\pi e(P+N)$, and working with the expression for mutual information,

$$I(X;Y) = h(Y) - h(Z)$$
$$\le \frac{1}{2}\log 2\pi e(P+N) - \frac{1}{2}\log 2\pi e N$$
$$= \frac{1}{2}\log\left(1+\frac{P}{N}\right), \tag{4.55}$$

with equality iff $X \sim \eta(0, P)$.

$$\therefore\ C = \max_{EX^2 \le P} I(X;Y) = \frac{1}{2}\log\left(1+\frac{P}{N}\right). \tag{4.56}$$

In order to give this expression an operational meaning in terms of the maximum rate that can be sent over a Gaussian channel with asymptotically small error probability, we need to show that this expression is the supremum (least upper bound or maximum if it exists) of the achievable rates for the channel. As before, we only outline the steps in the proof here.

To prove achievability, we need to show that for a Gaussian channel with a power constraint P there exists a sequence of $(2^{nR}, n)$ codes with codewords satisfying the power constraint such that the maximal probability of error $\lambda^{(n)}$ tends to zero. The capacity of the channel is the supremum (maximum) of the achievable rates.

We outline the steps in the proof as follows. As before for the discrete case, we use random codes and jointly typical decoding, only here we need to impose the average power constraint.

We generate elements of each codeword by letting $X_i(w), i = 1, 2, \ldots, n, w = 1, \ldots, 2^{nR}$ be iid $\sim \eta(0, P - \epsilon)$, and form codewords according to $X^n(1), X^n(2), \ldots, X^n(2^{nR}) \in \mathbb{R}^n$. The transmitter and receiver know the codebook.

To send the message w, the transmitter sends the wth codeword $X^n(w)$ in the codebook.

After passing through the additive Gaussian noise channel, Y^n is the input to the receiver. The receiver decodes Y^n as that $X^n(i)$ that is jointly typical with Y^n, which is represented by the event

$$E_i = \left\{ (X^n(i), Y^n) \in A_\epsilon^{(n)} \right\} .$$

If there is not just one $X^n(j)$ that is jointly typical with Y^n or the codeword does not satisfy the power constraint, the receiver declares an error, which is denoted by the event $\hat{W} \neq W$.

Using the Union bound [16], we can see that the error probability is upper bounded by the sum of three terms, the event where the generated codeword does not satisfy the power constraint, the event where codeword (say) $X^n(i)$ is sent but is not found to be jointly typical with the received Y^n, and the event that another codeword is found to be jointly typical with Y^n.

We can show that the term due to the codeword power not satisfying the power constraint approaches zero asymptotically from

$$\Pr \left\{ \left| \frac{1}{n} \sum_{i=1}^n X_i^2 - P \right| > \epsilon \right\} \leq \frac{\text{var} \left(\frac{1}{n} \sum X_i^2 \right)}{\epsilon}$$

which simplifies to

$$\Pr \left\{ \left| \frac{1}{n} \sum X_i^2 - P \right| > \epsilon \right\} \leq \frac{2P^2}{n\epsilon}$$

We can see that the second term in the union bound of the error probability is asymptotically small since if $X^n(i)$ is transmitted, we know by the AEP for joint typicality that the probability of the typical set approaches one as n gets large.

Finally, the third term would occur when $X^n(i)$ is transmitted but one of the other codewords is found to be jointly typical. Again by the joint AEP, this occurs with probability $P((X^n(j), Y^n) \in A_\epsilon^{(n)}) \leq 2^{-n(I(X;Y)-3\epsilon)}$ for each term j not equal to i. There are $2^{nR} - 1$ such terms so the probability of this third event is upper bounded by $2^{-n(I(X;Y))-R-3\epsilon}$.

We see that if $R < I(X;Y) - 3\epsilon$ for n sufficiently large, a good rate R code exists. To finish the proof, we choose a good codebook and delete the worst half of the codewords to obtain a code with low maximal probability of error.

This demonstrates achievability.

4.8   CONVERSE TO THE CODING THEOREM FOR GAUSSIAN CHANNELS

To prove the converse for our Gaussian channel with power constraint P, we must show that if $P_e^{(n)} \to 0$, then

$$R \le C = \frac{1}{2} \log \left(1 + \frac{P}{N} \right).$$

We start with the codebook designed as in the achievability proof [5]. Consider any $2^{nR}, n$ code that satisfies the power constraint, i.e.,

$$\frac{1}{n} \sum_{i=1}^{n} x_i^2(w) \le P$$

We also know that the data processing inequality holds for our coding/decoding method and channel, so $W \to X^n \to Y^n \to \hat{W}$. We assume a uniform distribution on W, and consider

$$nR = H(W) = I(W;Y^n) + H(W|Y^n) \le I(W;Y^n) + n\left(\frac{1}{n} + RP_e^{(n)} \right) \qquad (4.57)$$

where upon using Fano's inequality, we know that the term in parenthesis goes to 0 as $P_e^{(n)} \to 0$ for n sufficiently large.

Working with the first term on the right, we expand $I(W;Y^n)$ as

$$I(W;Y^n) \le I(X^n;Y^n)$$
$$= \sum_{i=1}^{n} h(Y_i) - \sum_{i=1}^{n} h(Z_i)$$
$$= \sum_{i=1}^{n} I(X_i;Y_i). \qquad (4.58)$$

where we have used in succession, the data processing inequality, the definition of mutual information, the independence of X and Z, the independence bound on the differential entropy $h(Y)$, the independence of the Z_i, and the definition of mutual information. Invoking the definition of the additive Gaussian noise channel and the power constraint on X_i, we know that

$$h(Y_i) \le \frac{1}{2} \log 2\pi e(P_i + N).$$

which we use in the prior sequence of inequalities to get

$$\sum_{i=1}^{n} (h(Y_i) - h(Z_i)) \le \sum_{i=1}^{n} \left[\frac{1}{2} \log 2\pi e(P_i + N) - \frac{1}{2} \log 2\pi e N \right]$$
$$\le \sum_{i=1}^{n} \frac{1}{2} \log \left(1 + \frac{P_i}{N} \right).$$

To simplify this last expression further, we note that since each of the codewords satisfies the power constraint, so does their average,

$$\frac{1}{n} \sum_{i=1}^{n} P_i \le P .$$

(4.59)

Then, noting that the function $f(x) = \frac{1}{2} \log(1 + x)$ is a concave function of x, so we apply Jensen's inequality (see appendices) to obtain

$$\frac{1}{2} \log \frac{1}{n} \sum_{i=1}^{n} \left(1 + \frac{P_i}{N}\right) \le \frac{1}{2} \log \left(1 + \frac{1}{n} \sum_{i=1}^{n} \frac{P_i}{N}\right)$$
$$\le \frac{1}{2} \log \left(1 + \frac{P}{N}\right) .$$

(4.60)

Thus,

$$R \le \frac{1}{2} \log \left(1 + \frac{P}{N}\right)$$

since the term from Fano's inequality $\to 0$ as $n \to \infty$, and we have the converse.

That is, if $P_e^{(n)} \to 0$, the rate R has to be less than channel capacity.

4.9 EXPRESSIONS FOR CAPACITY AND THE GAUSSIAN CHANNEL

We calculated the capacity of channels with continuous inputs and outputs when the noise is Gaussian earlier in this chapter. The following theorem expands the utility of the Gaussian result.

Theorem 4.14   *Consider the additive noise channel in Fig. 2.4, where $E[W^2] \le S$ and $var(\varsigma) = \sigma^2$. The capacity of this channel is bounded by*

$$\frac{1}{2} \log \left(1 + \frac{S}{\sigma^2}\right) \le C \le \frac{1}{2} \log \left[2\pi e \left(S + \sigma^2\right)\right] - h(\varsigma) .$$

In essence, this theorem says that for a fixed noise variance, Gaussian noise is the worst since it lower bounds the channel capacity.

Proof. We begin with the right inequality. With $E(W) = 0$ and $E(W^2) = S$, $X = W + \varsigma$, $E(\varsigma) = 0$, and $var(\varsigma) = \sigma^2$, $var(X) = S + \sigma^2$. Further, since

$$I(W; X) = h(X) - h(\varsigma) ,$$

(4.61)

and $h(\zeta)$ is independent of the input, then maximization of $I(W; X)$ is equivalent to maximization of $h(X)$. We know that this maximization is achieved by a zero mean, Gaussian random value with $\text{var}(X) = S + \sigma^2$, so

$$h(X) \le \frac{1}{2} \log 2\pi e (S + \sigma^2)$$

and therefore,

$$I(W; X) \le \frac{1}{2} \log 2\pi e (S + \sigma^2) - h(\zeta)$$

and the right inequality follows.

For the left inequality, let the input pdf be Gaussian,

$$f_W(w) = \frac{1}{\sqrt{2\pi S}} e^{-w^2/2S}, \quad -\infty < w < \infty,$$

and the pdf of ζ be zero mean, Gaussian,

$$f_\zeta(\zeta) = \frac{1}{\sqrt{2\pi\sigma}} e^{-\zeta^2/2\sigma^2}, \quad -\infty < \zeta < \infty,$$

so

$$f_X(x) = \frac{1}{\sqrt{2\pi(S + \sigma^2)}} e^{-x^2/2(S+\sigma^2)}, \quad -\infty < x < \infty.$$

Then from Eq. (4.61),

$$
\begin{aligned}
I(W; X) &= h(X) - h(\zeta) \\
&= \frac{1}{2} \log 2\pi e (S + \sigma^2) - \frac{1}{2} \log 2\pi e \sigma^2 \\
&= \frac{1}{2} \log \left(1 + \frac{S}{\sigma^2}\right).
\end{aligned}
$$

Now, consider arbitrary W and X that satisfy the given conditions,

$$
\begin{aligned}
-I(W; X) + \frac{1}{2} \log \left(1 + \frac{S}{\sigma^2}\right) &= - \int_{-\infty}^{\infty} \int_{-\infty}^{\infty} f_W(w) f_\zeta(x - w) \log \frac{f_\zeta(x - w)}{f_X(x)} \, dw \, dx \\
&\quad + \int_{-\infty}^{\infty} \int_{-\infty}^{\infty} f_W(w) f_\zeta(x - w) \frac{1}{2} \log \left(1 + \frac{S}{\sigma^2}\right) \, dw \, dx \\
&= \int_{-\infty}^{\infty} \int_{-\infty}^{\infty} f_W(w) f_\zeta(x - w) \log \left[\frac{f_\zeta(x - w) f_{S+\sigma^2}(x)}{f_X(x) f_{\sigma^2}(x - w)}\right]^{-1} \, dw \, dx
\end{aligned}
$$

where $f_{S+\sigma^2}(\cdot)$ and $f_{\sigma^2}(\cdot)$ denote Gaussian pdfs with the subscripted variances.

Using $\log \beta \leq (\log e)(\beta - 1)$,

$$I(W; X) - \frac{1}{2} \log \left(1 + \frac{S}{\sigma^2}\right) \leq \log_e \int_{-\infty}^{\infty} \int_{-\infty}^{\infty} f_W(w) f_\xi(x - w) \left\{ \frac{f_X(x) f_{\sigma^2}(x - w)}{f_\xi(x - w) f_{S+\sigma^2}(x)} - 1 \right\} dw \, dx$$

$$= \int_{-\infty}^{\infty} \int_{-\infty}^{\infty} \left\{ \frac{f_W(w) f_X(x) f_{\sigma^2}(x - w)}{f_{S+\sigma^2}(x)} \right\} dw \, dx - 1 \, .$$

Since $f_W(w)$ is Gaussian,

$$\int_{-\infty}^{\infty} f_W(w) f_{\sigma^2}(x - w) \, dw = f_{S+\sigma^2}(x) \, ,$$

and the double integral gives

$$\int_{-\infty}^{\infty} f_X(x) \, dx = 1 \, ,$$

so

$$-I(W; X) + \frac{1}{2} \log \left(1 + \frac{S}{\sigma^2}\right) \leq 0$$

or

$$\frac{1}{2} \log \left(1 + \frac{S}{\sigma^2}\right) \leq I(W; X) \, .$$

The result follows since C is the maximum (sup) over all input pdfs.   □

The basic expression for the capacity of an additive Gaussian noise channel is often further exploited to obtain very interesting and important results for more interesting channels. Parallel independent Gaussian channels with a common power constraint and channels with colored Gaussian noise play an important role in practical communications and compression problems. These channels are considered in the next two sections.

4.9.1   PARALLEL GAUSSIAN CHANNELS [4, 5]

Here we consider independent Gaussian channels in parallel with a common power constraint. The objective is to distribute the total power among the channels so as to maximize the capacity. This models a non-white additive Gaussian noise channel where each parallel component represents a different frequency.

For each parallel channel j,

$$Y_j = X_j + Z_j$$

with $Z_j \sim \eta(0, N_j)$ and the noise is assumed independent from channel to channel. The common power constraint is

$$E \sum_{i=1}^{k} X_i^2 \leq P \, . \tag{4.62}$$

We wish to distribute the power among the various channels so as to maximize the total capacity. The information capacity of the channel (the parallel channels) is

$$C = \max_{f(x_1,\ldots,x):\sum EX_i^2 \leq P} I(X_1,\ldots,X_k;Y_1,\ldots,Y_k) \,.$$

We calculate the distribution that achieves the information capacity for this channel (Note that the term Information Capacity corresponds to the expression for channel capacity based on the maximization of mutual information. This term is used for this quantity before a channel coding theorem is proved, thus giving the information capacity physical meaning).

Since $Z_1, Z_2, \ldots, Z_k$ are independent and independent of $X_1, \ldots, X_k$,

$$I(X_1,\ldots,X_k;Y_1,\ldots,Y_k) = h(Y_1,\ldots,Y_k) - h(Y_1,\ldots,Y_k|X_1,\ldots,X_k)$$

$$= h(Y_1,\ldots,Y_k) - \sum_{i=1}^{k} h(Z_i)$$

$$\leq \sum_{i=1}^{k} \left(h(Y_i) - \sum h(Z_i) \right)$$

$$\leq \sum_{i=1}^{k} \frac{1}{2} \log \left(1 + \frac{P_i}{N_i} \right) \,, \tag{4.63}$$

where $P_i = EX_i^2$ and $\sum_i P_i = P$. Equality is achieved by

$$(X_1,\ldots,X_k) \sim \eta \left(0, \begin{bmatrix} P_1 & 0 & \cdots & 0 \\ 0 & P_2 & \cdots & \cdots \\ \vdots & \vdots & \vdots & \vdots \\ 0 & \cdots & 0 & P_k \end{bmatrix} \right) \,.$$

The problem is thus reduced to finding the power allotment that maximizes the capacity subject to the constraint $\sum P_i = P$. Using Lagrange multipliers,

$$J(P_1,\ldots,P_k) = \sum \frac{1}{2} \log \left(1 + \frac{P_i}{N_i} \right) + \lambda \left(\sum P_i - P \right) \,.$$

Differentiating with respect to P_j,

$$\frac{\partial}{\partial P_j} J(P_1,\ldots,P_k) = \frac{1}{2} \frac{1}{\left(1 + \frac{P_j}{N_j} \right)} \cdot \left(\frac{1}{N_j} \right) + \lambda$$

$$= \frac{1}{2} \cdot \frac{1}{Pj + N_j} + \lambda = 0 \,.$$

Solving for P_j, $\frac{1}{P_j+N_j} = -2\lambda$ or $P_j + N_j = \frac{-1}{2\lambda}$ so $P_j = v - N_j$.

We must impose the conditions

$$\sum_i P_i = P \text{ and } P_i \geq 0$$

so

$$\sum P_j = \sum(v - N_j) = kv - \sum_{j=1}^{k} N_j = P$$

which yields

$$v = \frac{1}{k}\left(P + \sum_{j=1}^{k} N_j\right) = (P + N)_{\text{avg.}}$$

However, the P_i must be non-negative and a solution of this form may not exist. In this case we use the Kuhn–Tucker conditions to verify that the solution that maximizes capacity is

$$P_i = (v - N_i)^+ \tag{4.64}$$

where

$$(x)^+ = \left\{ \begin{array}{l} x, \; x \geq 0 \\ 0, \; x < 0 \end{array} \right\}$$

and where v is chosen so that

$$\sum_{i=1}^{k}(v - N_i)^+ = P . \tag{4.65}$$

Figure 4.3 illustrates the parallel Gaussian water filling problem. The Kuhn–Tucker conditions imply that

$$\frac{1}{2}\frac{1}{P_j + N_j} + \lambda = \Lambda \text{ for } P_j > 0$$

or

$$P_j + N_j = \frac{1}{2(\Lambda - \lambda)} \text{ for } P_j > 0 .$$

But

$$\frac{1}{2}\frac{1}{P_j + N_j} + \lambda \leq \Lambda \text{ for } P_j = 0$$

or

$$N_j \geq \frac{1}{2(\Lambda - \lambda)} .$$

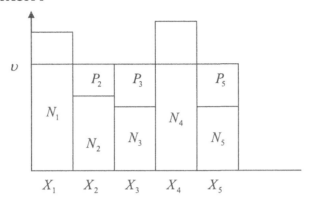

Figure 4.3: Water filling for parallel Gaussian channels.

The implication of the above result is that to maximize capacity, the power should be allocated to less-noisy channels first and then successively to other parallel channels. If a channel is too noisy, then we do not transmit information, that is we do not allocate any power, to this channel at all. These results, which are obtained using what are by now well-worn techniques (and the reader will find these techniques useful in many problems), can be seen to have applicability to communications systems using OFDM or DSL. Indeed, this result was the original motivation for DSL. This result can be seen to be an illustration of what is called the water filling approach, since transmitted power is allocated to channels until all channels have a noise plus transmitted power amount at the constraint level.

4.9.2 CHANNELS WITH COLORED GAUSSIAN NOISE [4, 5]

We consider parallel Gaussian channels where the noise is dependent. For channels with memory, we can consider a block of n consecutive uses of the channels as n channels in parallel with dependent noise. We develop an expression for capacity, sometimes called the information capacity, but do not prove a coding theorem to establish its physical interpretation as channel capacity. These proofs are left to the references.

Let K_z and K_x be the covariance matrices of the noise and the input, respectively. The input for constraint is

$$\frac{1}{n} \sum_i EX_i^2 \le P \tag{4.66}$$

or equivalently

$$\frac{1}{n} \operatorname{tr}(K_x) \le P . \tag{4.67}$$

The power constraint depends on n so the capacity must be calculated for each n. We can write

$$I(X_1, X_2, \ldots, X_n; Y_1, \ldots, Y_n) = h(Y_1, Y_2, \ldots, Y_n) - h(Z_1, Z_2, \ldots, Z_n) . \tag{4.68}$$

$\{Z_i\}$ is not dependent on $\{X_i\}$, so finding capacity amounts to maximizing $h(Y_1, \ldots, Y_n)$. This is maximum for Gaussian Y which is achieved when X is Gaussian. Thus, since the input and noise are independent, $K_y = K_x + K_z$ and

$$h(Y_1, \ldots, Y_n) = \frac{1}{2} \log \left[(2\pi e)^n | K_x + K_z| \right] . \tag{4.69}$$

The problem is thus reduced to choosing K_x to maximize $|K_x + K_z|$, subject to the power constraint $-\text{tr}(K_x)$ constraint. Write

$$K_z = Q \Lambda Q^T \quad \text{when} \quad Q Q^T = I .$$

Then

$$\begin{aligned} |K_x + K_z| &= \left| K_x + Q \Lambda Q^T \right| \\ &= |Q| \left| Q^T K_x Q + \Lambda \right| \left| Q^T \right| \\ &= \left| Q^T K_x Q + \Lambda \right| = |A + \Lambda| \end{aligned} \tag{4.70}$$

where $A = Q^T K_x Q$. For any matrices C and B,

$$\text{tr}(BC) = \text{tr}(CB) , \tag{4.71}$$

so

$$\begin{aligned} \text{tr}(A) = \text{tr } Q^T K_x Q &= \text{tr } \left(Q Q^T K_x \right) \\ &= \text{tr } K_x \end{aligned}$$

and we can maximize $|A + \Lambda|$ subject to trace constraint on K_x.

By Hadamard's inequality [37, p. 233],

$$|K| \leq \Pi_i K_{ii}$$

with equality iff K is diagonal, and also we have

$$|A + \Lambda| \leq \Pi_i (A_{ii} + \lambda_i) \tag{4.72}$$

with equality iff A is diagonal. A is also subject to the constraints $\frac{1}{n} \sum_i A_{ii} \leq P$ and $A_{ii} \geq 0$.

Using the Kuhn–Tucker conditions on

$$J = h(Y_1, \ldots, Y_n) + \gamma \left(\frac{1}{n} \sum A_{ii} - P \right)$$

so

$$\frac{1}{[A_{jj} + \lambda_j]} = 2 \left[\Gamma - \frac{\gamma}{n} \right]$$

or

$$[A_{jj} + \lambda_j] = \frac{1}{2 \left[\Gamma - \frac{\gamma}{n} \right]} = \nu$$

for $A_{jj} > 0$,
and

$$\frac{1}{\lambda_j} \leq 2 \left[\Gamma - \frac{\gamma}{n} \right]$$

or

$$\lambda_j \geq \frac{1}{2 \left[\Gamma - \frac{\gamma}{n} \right]} = \nu \ \text{ for } \ A_{jj} = 0 .$$

This says pick $A_{jj} = 0$ if $\lambda_j > \nu$, otherwise choose A_{jj} to satisfy $A_{jj} + \lambda_j = \nu$. These conditions can be expressed as

$$A_{jj} = [\nu - \lambda_j]^+ .$$

Choose ν to satisfy $\frac{1}{n} \sum_i A_{ii} = P$ or

$$\frac{1}{n} \sum_i A_{ii} = \frac{1}{n} \left[n\nu - \sum \lambda_i \right] = P$$

or

$$P = \nu - \frac{1}{n} \sum \lambda_i .$$

The simple independent parallel Gaussian channel result has now been extended to the even more interesting case of parallel Gaussian channels where the noise is dependent. In many applications when we do not know the distribution of the channel noise, we can still use this result to obtain substantive insights into a practical coding approach.

4.10   BAND-LIMITED CHANNELS

Now we consider the case of a channel with bandwidth W Hz and additive Gaussian noise [1, 4, 5]. We can represent the input and output by samples taken $1/2W$ secs. apart, and the input samples are corrupted by additive white Gaussian noise with power spectral density $\frac{N_0}{2}$. The noise samples can be shown to be iid Gaussian with zero mean and variance $\frac{N_0}{2}$.

We can now use our prior expression for capacity for this situation. In particular, we have that the definition of information capacity for an average power constraint P is

$$C = \max_{p(x):EX^2 \leq P} I(X;Y) \tag{4.73}$$

which we have shown for the Gaussian channel to simplify to

$$C = \frac{1}{2} \log\left(1 + \frac{P}{N}\right) \tag{4.74}$$

and is attained for $X \sim \eta(0, P)$, where $P = EX^2$.

For a band-limited Gaussian channel with bandwidth= W Hz, used over a time interval T, we have $2W$ samples/sec. or $2WT$ samples in a T sec. interval. Then,

$$P_{av} = \frac{1}{T} \int_0^T EX^2(t)\,dt = \frac{1}{T} \sum_{i=1}^{2WT} EX_i^2$$

(the average power over the interval T)

$$= \frac{1}{T} 2WTP = 2WP$$

$$\therefore P = \frac{P_{av}}{2W}.$$

Further, we see that the total noise power in bandwidth W Hz and time interval T is

$$\left(\frac{N_0}{2}\right) \qquad 2W(T) \qquad = N_0 WT$$
$$\text{(watts/Hz)} \quad \text{(Hz) (sec.)} \quad = \text{watt-secs.}$$

and there are $2WT$ noise samples, so the noise power/sample is $\frac{N_0 WT}{2WT} = \frac{N_0}{2}$. Thus, using (4.74),

$$C = \frac{1}{2} \log\left(1 + \frac{P_{av}/2W}{N_0/2}\right)$$

$$= \frac{1}{2} \log(1 + \frac{P_{av}}{N_0 W}) \text{ bits/sample}$$

or

$$C = W \log\left(1 + \frac{P_{av}}{N_0 W}\right) \text{ bits/sec.} \tag{4.75}$$

This capacity expression is widely known and has played a role in estimating the capacity of many channels, even those wherein the dominant noise is known not to be Gaussian, such as for estimating the capacity of the analog telephone channel for early telephone bandwidth modems.

4.11   NOTES AND ADDITIONAL REFERENCES

The content of this chapter is fairly conventional, although the ordering of the topics by section is slightly different than elsewhere. The primary influences on this chapter are the books by Gallager [4] and Cover and Thomas [5] and the paper by Shannon [29].

CHAPTER 5

Rate Distortion Theory and Lossy Source Coding

5.1 THE RATE DISTORTION FUNCTION FOR DISCRETE MEMORYLESS SOURCES

In Chapter 3 the transmitted data rate required to produce a discrete-time, discrete-amplitude source *exactly* (with no error) is considered, and the minimum rate necessary is shown to be the absolute or discrete entropy [5, 7]. The process of exactly representing a discrete-amplitude source with a reduced or minimum number of binary digits is called *noiseless source coding*, or *lossless source coding*. If the source to be transmitted is a continuous-amplitude random variable or random process, the source has an infinite number of possible amplitudes, and hence the number of bits required to reproduce the source *exactly* at the receiver is infinite. This is indicated by the fact that continuous-amplitude sources have infinite *absolute* entropy. Therefore, to represent continuous-amplitude sources in terms of a finite number of bits/source letter, we must accept the inevitability of some amount of reconstruction error or distortion.

We are thus led to the problem of representing a source with a minimum number of bits/source letter subject to a constraint on allowable distortion. This problem is usually called *source coding with a fidelity criterion* or *lossy source coding* Source coding with respect to some distortion measure may also be necessary for a discrete-amplitude source. For instance, if we are given a DMS with entropy H such that $H > C$, it may be necessary to accept some amount of distortion in the reproduced version of the DMS in order to reduce the required number of bits/source letter below C.

For source coding with a fidelity criterion, the function of interest is no longer H, but the rate distortion function, denoted $R(D)$. The rate distortion function $R(D)$ with respect to a fidelity criterion is the minimum information rate necessary to represent the source with an average distortion less than or equal to D. It is critical to observe that we must have a good source model and a distortion measure that is meaningful for the problem at hand. Both of these assignments have been and continue to be the principal challenges to rate distortion theory having a dramatic impact on the field of lossy source coding.

To be more specific, we again focus on the block diagram in Fig. 3.1, where the channel is ideal and the channel encoder/decoder blocks are identities, and on discrete memoryless sources. We must choose or be given a meaningful measure of distortion for the source/user pair.

If the source generates the output letter u and this letter is reproduced at the source decoder output as z, we denote the distortion incurred by this reproduction as $d(u, z)$. The quantity $d(u, z)$ is sometimes called a *single-letter* distortion measure or fidelity criterion. The average value of $d(\cdot, \cdot)$ over all possible Source outputs u and User inputs z is

$$\bar{d}\left(P_{Z|U}\right) = \sum_{u=1}^{J} \sum_{z=1}^{K} P_U(u) P_{Z|U}(z|u) d(u, z) , \qquad (5.1)$$

The average distortion in Eq. (5.1) is a function of the transition probabilities $\{P_{Z|U}(z|u), u = 1, 2, \ldots, J, z = 1, 2, \ldots, K\}$, which are determined by the source encoder/decoder pair. To find the rate distortion function, we wish only to consider those conditional probability assignments $\{P_{Z|U}\}$ that yield an average distortion less than or equal to some acceptable value D, called D-*admissible* transition probabilities, denoted by

$$\mathcal{P}_D = \left\{P_{Z|U}(z|u) : \bar{d}\left(P_{Z|U}\right) \leq D\right\} . \qquad (5.2)$$

For each set of transition probabilities, we have a mutual information

$$I(U; Z) = \sum_{u=1}^{J} \sum_{z=1}^{K} P_U(u) P_{Z|U}(z|u) \log \frac{P_{Z|U}(z|u)}{P_Z(z)} . \qquad (5.3)$$

We are now able to define the rate distortion function of the source with respect to the fidelity criterion $d(\cdot, \cdot)$ as

$$R(D) = \min_{P_{Z|U} \in \mathcal{P}_D} I(U; Z) \qquad (5.4)$$

for a chosen or given fixed value D. The importance of the rate distortion function is attested to by the fact that for a channel of capacity C, it is possible to reproduce the source at the receiver with an average distortion D if and only if $R(D) \leq C$.

A useful property of $R(D)$ is stated in the following theorem.

Theorem 5.1   *For a DMS with J output letters,*

$$0 \leq R(D) \leq \log J . \qquad (5.5)$$

Proof. Straightforwardly using basic definitions and properties,

$$\begin{aligned} 0 \leq I(U; Z) &= H(U) - H(U|Z) \\ &\leq H(U) \leq \log J . \end{aligned} \qquad (5.6)$$

Now, $R(D)$ is the minimum of $I(U; Z)$ over the admissible conditional probabilities; hence Eq. (5.5) follows. $\square$

The evaluation of the rate distortion function is not straightforward, even for discrete memoryless sources. The following simple example illustrates a particular approach to developing rate distortion functions that has often been useful.

Example 5.2 [7]   Here we examine a special case of a DMS called a binary symmetric source (BSS) that produces a 0 with probability p and a 1 with probability $1 - p$. If we let the source and reconstruction alphabets be 0,1, the single-letter distortion measure is specified to be

$$d\,(u, z) = \begin{cases} 0, & u = z \\ 1, & u \neq z \,. \end{cases} \tag{5.7}$$

Then, for $p \leq \frac{1}{2}$,

$$\begin{aligned} R(D) = &- p \log p - (1 - p) \log(1 - p) + D \log D \\ &+ (1 - D) \log(1 - D), \qquad 0 \leq D \leq p \,. \end{aligned} \tag{5.8}$$

This $R(D)$ is plotted in Fig. 5.1 for $p = 0.1, 0.2, 0.3$, and 0.5. The reader should verify that $R(D) \leq H(U)$ for each p and that $R(p) = 0$.

This BSS is often called a Bernoulli(p) source and the result can be stated more generally as

Theorem 5.3   *The rate distortion function for a Bernoulli(p) source with Hamming distortion is given by*

$$R(D) = \begin{cases} H(p) - H(D), & 0 \leq D \leq \min\{p, 1 - p\}\,, \\ 0, & D > \min\{p, 1 - p\}\,. \end{cases} \tag{5.9}$$

Proof. For the Bernoulli(p) source with a Hamming distortion measure, letting $p < \frac{1}{2}$, we wish to find an expression for the rate distortion function,

$$R(D) = \min_{p(\hat{x}|x):\sum_{(x,\hat{x})} p(x)p(\hat{x}|x)d(x,\hat{x}) \leq D} I(X; \hat{X}) \,. \tag{5.10}$$

To develop an expression for the distortion, we let $\oplus$ denote modulo 2 addition, so $X \oplus \hat{X} = 1$ is equivalent to $X \neq \hat{X}$. Berger provides a direct solution to the rate distortion problem using a parametric approach, which is beyond the scope of this book. Here, rather than attempt to minimize $I(X; \hat{X})$ directly, instead, we use an approach due to Shannon, wherein we find a lower bound and then show that this lower bound is achievable.

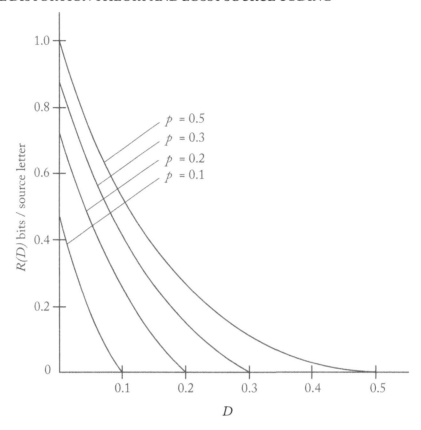

Figure 5.1: $R(D)$ for a BSS with $p \leq \frac{1}{2}$ (Ex. 5.2). From [7].

For any joint distribution satisfying the distortion constraint, we have

$$I(X; \hat{X}) = H(X) - H(X|\hat{X}) \tag{5.11}$$
$$= H(p) - H(X \oplus \hat{X}|\hat{X}) \tag{5.12}$$
$$\geq H(p) - H(X \oplus \hat{X}) \tag{5.13}$$

since conditioning cannot increase entropy, then

$$I(X; \hat{X}) \geq H(p) - H(D), \tag{5.14}$$
$$\tag{5.15}$$

since $\Pr(X \neq \hat{X}) \leq D$ and $H(D)$ increases with D for $D \leq \frac{1}{2}$. Thus

$$R(D) \geq H(p) - H(D) . \tag{5.16}$$

To show that the lower bound is achieved, we need the optimum backward channel result from the Shannon lower bound, and so we defer showing equality is achieved until the later section on the Shannon lower bound.

5.2   THE RATE DISTORTION FUNCTION FOR CONTINUOUS AMPLITUDE SOURCES

For a discrete-time continuous amplitude source with single-letter distortion measure $d(u, z)$ [5, 7], each conditional pdf relating the source output to the user input produces an average distortion given by

$$\bar{d}\left(f_{Z|U}\right) = \int_{-\infty}^{\infty} \int_{-\infty}^{\infty} f_U(u) f_{Z|U}(z|u) d(u, z) du \, dz \tag{5.17}$$

and a mutual information

$$I(U; Z) = \int_{-\infty}^{\infty} \int_{-\infty}^{\infty} f_U(u) f_{Z|U}(z|u) \log \frac{f_{Z|U}(z|u)}{f_Z(z)} du \, dz \ . \tag{5.18}$$

The admissible pdfs are described by the set

$$\mathcal{P}_D = \left\{ f_{Z|U}(z|u) : \bar{d}\left(f_{Z|U}\right) \leq D \right\} \ .$$

The rate distortion function is then defined as[1]

$$R(D) = \min_{f_{Z|U} \in \mathcal{P}_D} I(U; Z) \ . \tag{5.19}$$

A significant difference between the rate distortion functions for discrete-amplitude and continuous-amplitude sources is that for $R(D)$ in Eq. (5.19), as $D \to 0$, $R(D) \to \infty$.   □

Analytical calculation of the rate distortion function for continuous-amplitude sources often is extremely difficult, and relatively few such calculations have been accomplished. We present the results of one such calculation in the following example, which uses an approach due to Shannon that is similar to that employed for the BSS in the prior section.

Example 5.4   For the squared-error distortion measure

$$d(u - z) = (u - z)^2 \ , \tag{5.20}$$

a discrete-time, memoryless Gaussian source with zero mean and variance σ^2 has the rate distortion function

$$R(D) = \begin{cases} \frac{1}{2} \log \frac{\sigma^2}{D}, & 0 \leq D \leq \sigma^2 \ , \\ 0, & D > \sigma^2 \ . \end{cases} \tag{5.21}$$

---

[1]Strictly speaking, the "min" in Eq. (5.19) should be replaced with "inf," denoting infimum or greatest lower bound.

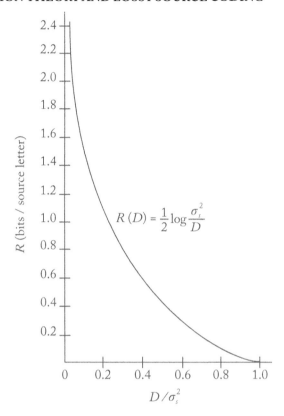

Figure 5.2: $R(D)$ for a memoryless Gaussian source and squared-error distortion (Ex. 5.4).

Proof. Let X be $\sim \mathcal{N}(0, \sigma^2)$. By the rate distortion theorem, we have

$$R(D) = \min_{f(\hat{x}|x):E(\hat{X}-X)^2 \leq D} I(X; \hat{X}) . \tag{5.22}$$

As in the previous example of a BSS with Hamming distortion, we first find a lower bound for the rate distortion function and then guess a distribution to show that the lower bound is achievable. We know that $E(X - \hat{X})^2 \leq D$, so we have the following

$$I(X; \hat{X}) = h(X) - h(X|\hat{X}) \tag{5.23}$$

$$= \frac{1}{2} \log(2\pi e)\sigma^2 - h(X - \hat{X}|\hat{X}) \tag{5.24}$$

$$\geq \frac{1}{2} \log(2\pi e)\sigma^2 - h(X - \hat{X}) \tag{5.25}$$

$$\tag{5.26}$$

since conditioning cannot increase entropy, and continuing

$$\geq \frac{1}{2} \log(2\pi e)\sigma^2 - h(\mathcal{N}(0, E(X - \hat{X})^2)) \tag{5.27}$$
$$\tag{5.28}$$

because the Gaussian distribution maximizes entropy,

$$= \frac{1}{2} \log(2\pi e)\sigma^2 - \frac{1}{2} \log(2\pi e)E(X - \hat{X})^2 \tag{5.29}$$
$$\geq \frac{1}{2} \log(2\pi e)\sigma^2 - \frac{1}{2} \log(2\pi e)D \tag{5.30}$$
$$\tag{5.31}$$

invoking the condition that $E(X - \hat{X})^2 \leq D$,

$$= \frac{1}{2} \log \frac{\sigma^2}{D}, \tag{5.32}$$

where (5.25) follows from the fact that conditioning reduces entropy and (5.27) follows from the fact that the normal distribution maximizes the entropy for a given second moment. Hence

$$R(D) \geq \frac{1}{2} \log \frac{\sigma^2}{D}. \tag{5.33}$$

Therefore, the rate distortion function for the Gaussian source with squared error distortion is

$$R(D) = \begin{cases} \frac{1}{2} \log \frac{\sigma^2}{D}, & 0 \leq D \leq \sigma^2, \\ 0, & D > \sigma^2. \end{cases} \tag{5.34}$$

as illustrated in Fig. 5.2.   □

Equation (5.34) is a relatively simple result that is given added importance by the fact that the Gaussian source is a worst-case source in the sense that it requires the maximum rate of all possible sources to achieve a specified mean square-error distortion. Specifically, any memoryless, zero-mean, continuous-amplitude source with variance σ_s^2 has a rate distortion function $R(D)$ with respect to the squared-error distortion measure that is upper bounded as

$$R(D) \leq \frac{1}{2} \log \frac{\sigma_s^2}{D}, \qquad 0 \leq D \leq \sigma_s^2. \tag{5.35}$$

This result implies that if the Gaussian source with the chosen average distortion can be transmitted over the channel, then any other memoryless source with the same variance will be supported by the same channel with the same average distortion. However, it may be possible to transmit at a lower rate for the given source if it is non-Gaussian. Of course, to find the required rate, the true source distribution must be known and an expression for its rate distortion function must be found. These conditions are not always easy to satisfy, so the Gaussian result is very useful.

We can rewrite (5.34) to express the distortion in terms of the rate,

$$D(R) = \sigma^2 2^{-2R} . \tag{5.36}$$

The approach to compression represented by $D(R)$ should be contrasted with the approach used to develop $R(D)$. The latter minimizes the rate for a given (chosen) distortion constraint, while the former implies minimizing the distortion subject to a rate constraint, that is,

$$D(R) = \min_{f(\hat{x}|x):I(X;\hat{X})\leq R} E(\hat{X} - X)^2 . \tag{5.37}$$

The $R(D)$ approach fits problems where a known distortion is acceptable and we need to minimize the rate. This approach fits audio coding and video coding where near-transparent quality is desired and the rate needed to accomplish this goal is accepted by the application. The $D(R)$ approach fits many network or wireless communications problems where the maximum rate available is fixed and the compression is accomplished to minimize the distortion.

5.3   THE SHANNON LOWER BOUND AND THE OPTIMUM BACKWARD CHANNEL

While it is difficult to obtain closed form solutions for $R(D)$ in general, there have been a number of lower bounds that have been developed [5, 7]. The most useful for us and for many applications is the Shannon lower bound (SLB) for difference distortion measures derived by Shannon in his 1959 paper [3]. For the discrete case, Shannon provided the following proof.

For a source X and its reconstruction $\hat{X}$ that achieves an average distortion $Ed(X, \hat{X}) \leq D$ for some difference distortion measure, we expand $I(X; \hat{X})$ as

$$
\begin{aligned}
I(X; \hat{X}) &= H(X) - H(X|\hat{X}) \\
&= H(X) - \sum_{\hat{x}} p(\hat{x})H(X|\hat{X} = \hat{x}) \\
&\geq H(X) - \sum_{\hat{x}} p(\hat{x})\phi(D_{\hat{x}})
\end{aligned}
$$

where we have used the notation $\phi(D_{\hat{x}})$ to denote the maximum of the conditional entropy $H(X|\hat{X} = \hat{x})$ that achieves $D_{\hat{x}} = \sum_x p(x|\hat{x})d(x, \hat{x})$. We next use Jensen's inequality to obtain

$$
\begin{aligned}
I(X; \hat{X}) &\geq H(X) - \phi\left(\sum_{\hat{x}} p(\hat{x})D_{\hat{x}}\right) \\
&\geq H(X) - \phi(D) ,
\end{aligned}
$$

where the last expression follows since $D_{\hat{x}} \leq D$.

These are basically the set of steps we followed in developing the lower bounds on $R(D)$ for the BSS and the memoryless Gaussian source in the preceding section.

To find the condition where the lower bound produced here and the bounds obtained for these specific sources is satisfied with equality, Shannon showed that for a continuous source X and a difference distortion measure $d(x, \hat{x}) = d(x - \hat{x})$ subject to the constraint

$$\iint p(x)p(\hat{x}|x)d(x - \hat{x}) \, dx \, d\hat{x} \leq D .$$

then

$$R(D) \geq h(X) + \int g(e) \log g(e) \, de$$

or

$$R(D) \geq h(X) - \max h(g(e)),$$

where $g(e) = g(x - \hat{x})$ and the maximum is taken over all probability densities for the error that satisfy the average distortion constraint [7]. Equality is achieved iff

$$\int p(\hat{x})g(x - \hat{x}) \, d\hat{x} = p(x) .$$

The equality condition produces the Shannon optimum backward channel, which states that the reconstructed value and the encoding error are statistically independent and sum to the input source value. It is the statistical independence result via the Shannon optimum backward channel that allows us to obtain the final form of $R(D)$ for the BSS and the memoryless Gaussian source in the following sections.

5.3.1   BINARY SYMMETRIC SOURCE

We can follow the recipe indicated by the Shannon backward channel condition to find a distribution that achieves equality in the lower bound on $R(D)$ for the BSS that we obtained in the preceding section. Specifically for the BSS, we find a conditional probability density of the source input X given the reconstructed source $\hat{X}$ when $\Pr(X \neq \hat{X}) = D$. Thus, we can write

$$P(\hat{X} = 0)(1 - D) + (1 - P(\hat{X} = 0))D = P(X = 0) = p , \tag{5.38}$$

Solving this equation for $P(\hat{X} = 0)$, we obtain

$$P(\hat{X} = 0) = \frac{p - D}{1 - 2D} . \tag{5.39}$$

which for $D \leq p \leq \frac{1}{2}$ yields a valid probability assignment for $\hat{X}$. It follows that

$$I(X; \hat{X}) = H(X) - H(X|\hat{X}) = H(p) - H(D) , \tag{5.40}$$

with expected distortion $P(X \neq \hat{X}) = D$. If $D \geq p$, then we can achieve $R(D) = 0$ by letting $\hat{X} = 0$ with probability 1.

We have therefore found an achievable probability assignment on the lower bound, and the rate distortion function for a binary source is

$$R(D) = \begin{cases} H(p) - H(D), & 0 \leq D \leq \min\{p, 1-p\}, \\ 0, & D > \min\{p, 1-p\}. \end{cases} \tag{5.41}$$

This function is illustrated in Fig. 5.1.

5.3.2 GAUSSIAN SOURCE

From the equality condition in the Shannon lower bound for difference distortion measures, we know that the reconstructed value, the source, and the encoding error must satisfy the optimum Shannon backward channel result, $X = \hat{X} + Z$, where $\hat{X}$ and Z are statistically independent. Since the source X is Gaussian with mean zero and variance σ^2 and the variance of the error is D, and the probability density for the error that maximizes the entropy is Gaussian, we find that $Z \sim \mathcal{N}(0, D)$ and $\quad \hat{X} \sim \mathcal{N}(0, \sigma^2 - D)$.

We have directly for these distributions that

$$I(X; \hat{X}) = \frac{1}{2} \log \frac{\sigma^2}{D}, \tag{5.42}$$

for $E(X - \hat{X})^2 = D$ and $D \leq \sigma^2$, and $R(D) = 0$ if $D > \sigma^2$, thus achieving the bound in (5.33).

The rate distortion function of a vector X of independent (but not identically distributed) Gaussian sources, as represented by Fig. 5.3, is calculated by the reverse water-filling theorem. This theorem says that we should encode the independent subsources X_i with equal distortion level λ, as long as λ does not exceed the variance of the transmitted subsource, and that one should not transmit at all those subsources whose variance is less than the distortion λ.

Theorem 5.5   (Reverse water-filling theorem) *For a vector X of independent random variables $X_1, X_2, ..., X_n$ such that $X_i \sim N(0, \sigma_i^2)$ and the distortion measure $D(\underline{X}, \hat{\underline{X}}) = E\left[\sum_{i=1}^{n}(X_i - \hat{X}_i)^2 = \sum_{i=1}^{n} D_i\right] \leq D$, the rate distortion function is*

$$R(D) = \min_{p(\hat{\underline{x}}|x): D(\underline{X}, \hat{\underline{X}}) \leq D} I(X; \hat{X}) = \sum_{i=1}^{n} \frac{1}{2} \log \frac{\sigma_i^2}{D_i}, \tag{5.43}$$

where

$$D_i = \begin{cases} \lambda & 0 \leq \lambda \leq \sigma_i^2 \\ \sigma_i^2 & \lambda > \sigma_i^2 \end{cases}. \tag{5.44}$$

To begin, we use standard arguments to simplify the mutual information of the n-vectors as

$$I(X; \hat{X}) = h(X) - h(X|\hat{X}) \tag{5.45}$$

$$= \sum_{i=1}^{n} h(X_i) - \sum_{i=1}^{n} h(X_i|X^{i-1}, \hat{X}) \tag{5.46}$$

$$\geq \sum_{i=1}^{n} h(X_i) - \sum_{i=1}^{n} h(X_i|\hat{X}_i) \tag{5.47}$$

$$= \sum_{i=1}^{n} I(X_i; \hat{X}_i) \tag{5.48}$$

$$\geq \sum_{i=1}^{n} R(D_i) \tag{5.49}$$

$$= \sum_{i=1}^{n} \left(\frac{1}{2} \log \frac{\sigma_i^2}{D_i} \right)^{+}, \tag{5.50}$$

where the superscript $+$ indicates that the term is zero if the quantity is negative.

From the Shannon backward channel result, we can achieve equality in the lower bound by choosing $f(x^n|\hat{x}^n) = \prod_{i=1}^{n} f(x_i|\hat{x}_i)$ and setting the distribution of each $\hat{X}_i \sim \mathcal{N}(0, \sigma_i^2 - D_i)$.

To find the rate distortion function for the n-vector X, we now have the optimization problem

$$R(D) = \min_{\sum D_i = D} \sum_{i=1}^{n} \max \left\{ \frac{1}{2} \ln \frac{\sigma_i^2}{D_i}, 0 \right\}. \tag{5.51}$$

To apply the Kuhn-Tucker conditions we use Lagrange multipliers and form the functional

$$J(D) = \sum_{i=1}^{m} \frac{1}{2} \ln \frac{\sigma_i^2}{D_i} + \lambda \sum_{i=1}^{m} D_i, \tag{5.52}$$

which upon differentiating with respect to D_i, we get

$$\frac{\partial J}{\partial D_i} = -\frac{1}{2} \frac{1}{D_i} + \lambda, \tag{5.53}$$

where λ is chosen so that

$$\frac{\partial J}{\partial D_i} \begin{cases} = 0, & \text{if } D_i < \sigma_i^2, \\ \leq 0, & \text{if } D_i \geq \sigma_i^2. \end{cases} \tag{5.54}$$

and $\sum_{i=1}^{m} D_i = D$.

Once we choose a constant λ, we code those subsources with variances greater than λ and spend zero bits on those subsources with a variance less than λ. This gives us the desired $R(D)$ for the n-vector X, which has the reverse water-filling interpretation.

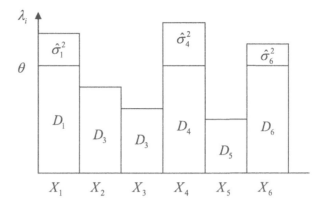

Figure 5.3: Reverse water-filling for independent Gaussian random variables.

Reverse water-filling is a classical result in rate distortion theory and it plays a major role in many applications. The following section provides the road map for applying the reverse water-filling result for real sources such as voice, still image, and video sources.

This theorem is discussed further in Section 5.8.

5.4   STATIONARY GAUSSIAN SOURCES WITH MEMORY

In this section, we present a classic derivation recasting the rate distortion problem for stationary Gaussian sources into a form that is easy to use in many applications [31]. This result apparently first appeared in [31].

Let A be an unitary matrix denoting an orthonormal linear transformation from a vector of random variables $\underline{X}$ to another vector of random variables $\underline{\Theta}$ as

$$\underline{\Theta} = A\underline{X}, \ \hat{\underline{X}} = A^{-1}\hat{\underline{\Theta}} = A^T \hat{\underline{\Theta}}. \tag{5.55}$$

The following relations between $\underline{X}$ and $\underline{\Theta}$ can be derived:
Mean squared error:

$$
\begin{aligned}
D(\underline{X}, \hat{\underline{X}}) \quad &= \quad E[(\underline{X} - \hat{\underline{X}})^T (\underline{X} - \hat{\underline{X}})] \\
&= \quad E[(\underline{\Theta} - \hat{\underline{\Theta}})^T A^T A(\underline{\Theta} - \hat{\underline{\Theta}})] \\
&= \quad E[(\underline{\Theta} - \hat{\underline{\Theta}})^T (\underline{\Theta} - \hat{\underline{\Theta}})] \\
\text{orthonormal} \quad & \\
&= \quad D(\underline{\Theta}, \hat{\underline{\Theta}});
\end{aligned}
\tag{5.56}
$$

Mutual information:

$$I(\underline{X}; \hat{\underline{X}}) \underset{|A|\neq 0}{=} I(\underline{\Theta}; \hat{\underline{\Theta}}) \geq \sum_{i=1}^{n} I(\Theta_i; \hat{\Theta}_i), \tag{5.57}$$

with equality if and only if (iff) Θ_i's are independent.

As shown above, both the distortion (chosen here to be the summation of squared errors) and the mutual information of a random process $\underline{X}$ are equal to those of the unitary transform of the random process $\underline{\Theta}$, and therefore the rate distortion function of $\underline{X}$ equals the rate distortion function of $\underline{\Theta}$.

Therefore, to utilize the reverse-water-filling result, we need to find $\underline{X}$'s unitary transform $\underline{\Theta}$ with independent elements. This leads us quite naturally to the well known Karhunen Lòeve Transform (KLT), which is also called principal component analysis, to decorrelate the source $\underline{X}$ as follows.

With the covariance function of a stationary Gaussian source denoted by

$$\phi(n) = E[x_i x_{i+n}], \tag{5.58}$$

and with Φ_n representing the $n \times n$ covariance matrix of the source, with entries $\phi(n)$, then, $\Phi_n = \{\phi(|i - j|), i, j = 1, ..., n\}$. Denoting $\{\underline{\psi}_i, i = 1, ..., n\}$ as the normalized eigenvectors of Φ_n with corresponding eigenvalues $\{\lambda_i, i = 1, ..., n\}$, then

$$\Phi_n \underline{\psi}_i = \lambda_i \underline{\psi}_i, \tag{5.59}$$

and

$$\Phi_n = \Psi_n \Lambda \Psi_n^T, \tag{5.60}$$

where $\Psi_n = [\underline{\psi}_1, \underline{\psi}_2, ..., \underline{\psi}_n]$.

Since covariance matrices are symmetric, there always exists an eigenvalue decomposition of the covariance matrix with real eigenvalues, and furthermore, covariance matrices are positive semi-definite, therefore all their eigenvalues are non-negative, yielding

$$\underline{\Theta} = \Psi_n^T \underline{X}. \tag{5.61}$$

Thus, the rate distortion function of a stationary Gaussian source $\underline{X}$ with covariance matrix Φ_n can be computed as the rate distortion function of a stationary Gaussian source $\underline{\Theta}$, where $\underline{\Theta}$ has independent Gaussian elements, each of variance λ_i, which are eigenvalues of the covariance matrix Φ_n. The rate distortion function of $\underline{\Theta}$ is in turn solved by the reverse-water filling theorem.

5.5   THE RATE DISTORTION FUNCTION FOR A GAUSSIAN AUTOREGRESSIVE SOURCE

The linear prediction model, which is an autoregressive (AR) model, has played a major role in the design of leading narrowband and wideband voice codecs for decades, and continues to do so. Berger [7] and Gray [32], in separate contributions in the late 60's and early 70's, derived the rate distortion function for Gaussian autoregressive (AR) sources for the squared error distortion measure. Given the relevance of this source model to real applications, we state the basic result here in the following theorem [7]:

Theorem 5.6   *Let $\{X_t\}$ be an mth-order autoregressive source generated by an i.i.d. $N(0, \sigma^2)$ sequence $\{Z_t\}$ and the autoregression constants $a_1, \ldots, a_m$. Then the MSE rate distortion function of $\{X_t\}$ is given parametrically by*

$$D_\vartheta = \frac{1}{2\pi} \int_{-\pi}^{\pi} \min\left[\vartheta, \frac{1}{g(\omega)}\right] d\omega, \tag{5.62}$$

and

$$R(D_\vartheta) = \frac{1}{2\pi} \int_{-\pi}^{\pi} \max\left[0, \frac{1}{2} \log \frac{1}{\vartheta g(\omega)}\right] d\omega, \tag{5.63}$$

where

$$g(\omega) = \frac{1}{\sigma^2} \left| 1 + \sum_{k=1}^{m} a_k e^{-jk\omega} \right|^2. \tag{5.64}$$

The points on the rate distortion function are obtained as the parameter ϑ is varied from the minimum to the maximum of the power spectral density of the source. ϑ can be associated with a value of the average distortion, and as illustrated in Fig. 5.4, only the shape of the power spectral density, $\Phi(\omega)$, above the value of ϑ is reproduced at the corresponding distortion level. The reverse water-filling interpretation is clearly evident from the shaded region in the figure.

Specifically, ϑ is related to the average distortion through the slope of the rate distortion function at the point where the particular average distortion is achieved. The parametric equations to solve the rate distortion optimization problem are not presented here but are left to the references.

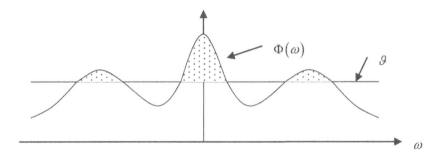

Figure 5.4: Example source, error, and reconstruction spectral densities.

The rate distortion function in this theorem offers a direct connection to many of the most widely used and standardized voice codecs. In the following section we present another key idea for applying rate distortion theory to practical sources such as voice and video.

5.6 COMPOSITE SOURCE MODELS AND CONDITIONAL RATE DISTORTION FUNCTIONS

The idea that sources may have multiple modes and can switch between modes probabilistically, was initially presented by Berger in his classic book, wherein he denoted such sources as composite sources[7] [33]. For a composite source, the choice of subsources is accomplished according to a probabilistic switch process, which is the side information Y [7, Sec. 6.1]. The power of such composite source models is that the individual subsources are able to capture local or fine dependence, while the switch process can represent changes that happen more globally as well as capturing model discontinuities. Given an appropriate number of carefully selected subsources and accurate switch modeling, time-varying or spatially-varying complex real world sources, such as voice and video, can be represented accurately.

A composite source with switch probability depending on the side information Y [7] is represented in Fig. 5.5.

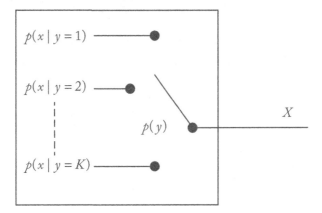

Figure 5.5: A composite source model with K subsources.

Motivated by the work of Berger [7] and others, research efforts explored, from the theoretical side, the properties of composite sources and also attempted to obtain expressions or bounds for the rate distortion functions of composite sources. The conditional rate distortion function is the rate of a source subject to a fidelity criterion when the encoder and decoder both have access to side information [33]. Thus, the conditional rate distortion function describes the rate required for a composite source subject to a fidelity criterion, where the side information is the switch process that selects the appropriate subsource at any time. The following definition of the conditional rate distortion function is from Gray [33].

Definition 5.7 The conditional rate distortion function of a source $\underline{X}$ with side information Y, which serves as the subsource information, is defined as

$$R_{\underline{X}|Y}(D) = \min_{p(\hat{\underline{x}}|\underline{x},y):D(\underline{X},\hat{\underline{X}}|Y)\leq D} I(\underline{X};\hat{\underline{X}}|Y), \tag{5.65}$$

where

$$D(\underline{X},\hat{\underline{X}}|Y) = \sum_{\underline{x},\hat{\underline{x}},y} p(\underline{x},\hat{\underline{x}},y)D(\underline{x},\hat{\underline{x}}|y),$$

$$I(\underline{X};\hat{\underline{X}}|Y) = \sum_{\underline{x},\hat{\underline{x}},y} p(\underline{x},\hat{\underline{x}},y)\log\frac{p(\underline{x},\hat{\underline{x}}|y)}{p(\underline{x}|y)p(\hat{\underline{x}}|y)}. \tag{5.66}$$

$R_{\underline{X}|Y}(D)$ is the lowest rate given that both encoder and decoder are allowed undistorted access to the switch process Y. It can be shown [33] that the conditional rate distortion function in Eq. (5.65) can also be expressed as

$$R_{\underline{X}|Y}(D) = \min_{D'_y s:D(\underline{X},\hat{\underline{X}}|Y)=\sum_y D_y\,p(y)\leq D} \sum_y R_{\underline{X}|y}(D_y)p(y), \tag{5.67}$$

and the minimum is achieved by adding up the individual, also called marginal, rate-distortion functions at points of equal slopes of the marginal rate distortion functions. Utilizing the prior expression for conditional rate distortion functions, the minimum is achieved at D_y's where the slopes $\frac{\partial R_{\underline{X}|Y=y}(D_y)}{\partial D_y}$ are equal for all y and $\sum_y D_y\,P[Y=y] = D$.

This conditional rate distortion function $R_{\underline{X}|Y}(D)$ can be used to write the following inequality involving the overall source rate distortion function $R_{\underline{X}}(D)$ [33]

$$R_{\underline{X}|Y}(D) \leq R_{\underline{X}}(D) \leq R_{\underline{X}|Y}(D) + I(\underline{X};Y), \tag{5.68}$$

where $I(\underline{X};Y)$ is the mutual information between $\underline{X}$ and Y and the equality in the leftmost inequality is achieved if and only if $\underline{X}$ and Y are independent. We can simplify the above and further bound $I(\underline{X};Y)$ by

$$I(\underline{X};Y) \leq H(Y) \leq \frac{1}{M}\log K, \tag{5.69}$$

where K is the number of subsources and M is the number of samples representing how often the subsources change. For many applications, such as for voice and video coding, K is much smaller than M so that the second term on the right involving the rate of sending the side information is negligible, and the rate distortion for the source is very close to the conditional rate distortion function.

5.7   THE RATE DISTORTION THEOREM FOR INDEPENDENT GAUSSIAN SOURCES–REVISITED

A mathematical characterization of the rate distortion function for Gaussian sources is given by the following theorem [5, 7].

Theorem 5.8   (Shannon's third theorem–Gaussian Sources) *For a vector X of independent random variables $X_1, X_2, ..., X_n$ and the distortion measure $D(\underline{X}, \hat{\underline{X}}) = E\left[\sum_{i=1}^{n}(X_i - \hat{X}_i)^2 = \sum_{i=1}^{n} D_i\right] \leq D$, the rate distortion function is*

$$R(D) = \min_{p(\hat{\underline{x}}|x):D(\underline{X},\hat{\underline{X}})\leq D} I(X;\hat{X}) = \sum_{i=1}^{n} \frac{1}{2} \log \frac{\sigma_i^2}{D_i}, \tag{5.70}$$

where $I(X;\hat{X})$ is the mutual information between X and $\hat{X}$. Given the source distribution $p(x)$,

the minimization is over all admissible test channels, that is, all $p(\hat{x}|x)$ satisfying the average distortion constraint.

The proof of achievability is long and fairly complicated. It is very similar to the proof of channel capacity given in Chapter 4 except that there is a condition added to joint typicality to include sequences that satisfy the distortion constraint, sometimes called distortion typical sequences. We only set up some notation here and leave the details of the proof to the references.

The proof, using a random coding argument and distortion typicality for encoding and decoding, involves showing there exists an encoder/decoder pair generated according to a test channel $p(\hat{x}|x)$ that achieves the rate distortion pair $(R(D), D)$. Notationally, the encoder and decoder are the functions f_n and g_n, respectively, and we get a length-n code with a codebook of 2^{nR} sequences. Instead of the achievability proof, we use a geometric sphere packing argument to motivate the achievability result.

We have a Gaussian source of variance σ^2 and use the encoder/decoder pair specified to get a $(2^{nR}, n)$ rate distortion code for this source with distortion D, which is a set of 2^{nR} codewords in $\mathbb{R}^n$ such that most source sequences of length n (all those that lie within a sphere of radius $\sqrt{n\sigma^2}$) are within a distance $\sqrt{nD}$ of some codeword. By the sphere packing argument, it is clear that the minimum number of codewords required is

$$2^{nR(D)} = \left(\frac{\sigma^2}{D}\right)^{n/2}. \tag{5.71}$$

The rate distortion theorem shows that this minimum rate is asymptotically achievable, i.e., that there exists a collection of spheres of radius $\sqrt{nD}$ that cover the space except for a set of arbitrarily small probability.

The proof of the converse to the Rate Distortion Theorem requires showing that the rate R of any rate distortion code exceeds the rate distortion function $R(D)$ evaluated at the distortion level $D = Ed(X^n, \hat{X}^n)$ achieved by the chosen code. The proof of the converse is outlined here, omitting steps involving well known equalities and inequalities used many times earlier in the book.

We have the encoding and decoding functions f_n and g_n defining a $(2^{nR}, n)$ rate distortion code and let the reconstructed value corresponding to X^n be $\hat{X}^n = \hat{X}^n(X^n) = g_n(f_n(X^n))$. We can then write

$$nR \geq H(\hat{X}^n) \tag{5.72}$$

$$\geq I(\hat{X}^n; X^n) \tag{5.73}$$

$$= \sum_{i=1}^{n} H(X_i) - \sum_{i=1}^{n} H(X_i | \hat{X}^n, X_{i-1}, \ldots, X_1) \tag{5.74}$$

$$\geq \sum_{i=1}^{n} I(X_i; \hat{X}_i) \tag{5.75}$$

$$\geq \sum_{i=1}^{n} R(Ed(X_i, \hat{X}_i)) \tag{5.76}$$

$$= n \sum_{i=1}^{n} \frac{1}{n} R(Ed(X_i, \hat{X}_i)) \tag{5.77}$$

$$\geq nR \left(\frac{1}{2} \sum_{i=1}^{n} Ed(X_i, \hat{X}_i) \right) \tag{5.78}$$

$$= nR(Ed(X^n, \hat{X}^n)) \tag{5.79}$$

$$= nR(D) , \tag{5.80}$$

This shows that $R(D)$ as defined by the minimum of the mutual information over the admissible set is the best that can be attained by any rate R code for the particular distortion.

5.8 APPLICATIONS OF $R(D)$ TO SCALAR QUANTIZATION

Perhaps the most familiar and most widely used lossy compression technique is scalar quantization and the corresponding code assignment [34]. We present three scalar quantization examples here for a zero mean, unit variance, memoryless Gaussian source and compare their performance to the rate distortion bound for a Gaussian source subject to the mean squared error distortion measure. First we consider a four-level quantizer designed for a Gaussian input where the step points and output levels are optimized to minimize the mean squared quantization error (MMSE), and the immediately following example considers an 8-level MMSE quantizer for the same Gaussian source. The third example evaluates an 8-level uniform quantizer for the same source. The rate required for the code assignment is compared by using the entropy of each quantizer output. (Rate and distortion must always be paired in any comparison.)

Example 5.9 A memoryless, zero-mean, unit-variance Gaussian source is quantized using a four-level MMSE Gaussian quantizer with the characteristic summarized in the Table 5.1.

Table 5.1:

i	x_i	y_i
1	-0.9816	-1.510
2	0.0	-0.4528
3	0.9816	0.4528
4	∞	1.510

$D = 0.1175$
$SNR = 9.30 \text{ dB}$

This process produces a discrete-time, discrete-amplitude memoryless source with output alphabet $u \in \mathcal{U} = \{-1.510, -0.4528, 0.4528, 1.510\}$. The probability assigned to each of these values must be calculated for the given Gaussian input and the quantizer step points in the table. Since all values of the source from $-\infty$ to -0.9816 are assigned to the output level $u = -1.510$, then

$$P_U(-1.510) = \int_{-\infty}^{-0.9816} \frac{1}{\sqrt{2\pi}} e^{-u^2/2} \, du = 0.1635 . \tag{5.81}$$

and similarly, we have

$$P_U(-0.4528) = \int_{-0.9816}^{0} \frac{1}{\sqrt{2\pi}} e^{-u^2/2} \, du = \frac{1}{2} - \int_{-\infty}^{-0.9816} \frac{1}{\sqrt{2\pi}} e^{-u^2/2} \, du$$

$$= 0.3365 \tag{5.82}$$

$$P_U(0.4528) = \int_{0}^{0.9816} \frac{1}{\sqrt{2\pi}} e^{-u^2/2} \, du = 0.3365 \tag{5.83}$$

and

$$P_U(1.510) = \int_{0.9816}^{\infty} \frac{1}{\sqrt{2\pi}} e^{-u^2/2} \, du = 0.1635 . \tag{5.84}$$

we can find the absolute entropy of this newly created DMS as 1.911 bits/source letter.

Three interesting observations can be made concerning these results. First, the quantizer has accomplished entropy reduction in that it has transformed a continuous amplitude source with infinite absolute entropy into a DMS with a finite entropy of 1.911 bits/source letter. Second, the minimum bit rate required to represent the quantizer outputs exactly is 1.911 bits/source letter, which does not seem to be significantly less than the 2 bits/source letter needed by a common

2-bits per level fixed length code. Hence, the extra effort required to achieve the minimum bit rate using Huffman or other lossless coding procedures may not be worthwhile, depending on the application. Third, from table, the mean-squared error distortion achieved by this quantizer is $D = 0.1175$ at a minimum rate of 1.911 bits/source letter. This result can be compared to the rate distortion function of a memoryless Gaussian source to ascertain how far away from optimum this quantizer operates.

Example 5.10   A memoryless, zero-mean, unit-variance Gaussian source is quantized using the 8-level MMSE Gaussian quantizer characteristic in the Table 5.2.

Table 5.2:

i	x_i	y_i
1	-0.748	-2.152
2	-1.510	-1.344
3	-0.5006	-0.7560
4	0.0	-0.2451
5	0.5006	0.2451
6	1.050	0.7560
7	1.748	1.344
8	∞	2.152

$D = 0.03454$
SNR $= 14.62$ dB

We find the equivalent DMS and its entropy, and then compare the performance of this quantizer to the rate distortion bound.

With $u = y$ in the table, it is evident that the quantizer is symmetric so we need only calculate four output probabilities explicitly as

$$P_U(-2.152) = \frac{1}{\sqrt{2\pi}} \int_{-\infty}^{-1.748} e^{-\gamma^2/2}\, d\gamma = 0.0401 = P_U(2.152)$$

$$P_U(-1.344) = \frac{1}{\sqrt{2\pi}} \int_{-1.748}^{-1.050} e^{-\gamma^2/2}\, d\gamma = 0.1068 = P_U(1.344)$$

$$P_U(-0.756) = \frac{1}{\sqrt{2\pi}} \int_{-1.050}^{-0.5006} e^{-\gamma^2/2}\, d\gamma = 0.1616 = P_U(0.756)$$

$$P_U(-0.2451) = \frac{1}{\sqrt{2\pi}} \int_{-0.5806}^{0} e^{-\gamma^2/2}\, d\gamma = 0.1915 = P_U(0.2451)\,.$$

with entropy

$$H(U) = 2.824 \text{ bits/letter}\,.$$

Since $D = 0.03454$ from the table, we can calculate the rate distortion bound from Eq. (5.35) as

$$R(D) = \frac{1}{2} \log \frac{\sigma_s^2}{D}, \quad 0 \le D \le \sigma_s^2$$

so

$$R(0.03454) = 2.428 \text{ bits/letter}\,.$$

Thus, the quantizer operates $2.824 - 2.428 = 0.396$ bits/letter above the rate distortion bound to achieve the same distortion, even if the quantizer output is entropy coded.

Example 5.11   Use the eight-level uniform quantizer in the associated table to quantize a memoryless, zero-mean, unit-variance Gaussian source. We find the output entropy and compare the performance of this quantizer to the rate distortion bound and to the quantizer in the previous example.

Table 5.3:

Number of Levels (L)	Step Size (Δ_{opt})	Minimum Mean Squared Error (D)
4	0.9957	0.1188
8	0.5860	0.03744
16	0.3352	0.01154

From the Table 5.3, the quantizer step size is ($\Delta = 0.586$) and so the output alphabet is $u \in \mathcal{U} = -2.051, -1.465, -0.879, -0.293, 0.293, 0.879, 1.465, 2.051$ with probabilities

$$P_U(-2.051) = \frac{1}{\sqrt{2\pi}} \int_{-\infty}^{-1.758} e^{-\gamma^2/2} \, d\gamma = 0.0392 = P_U(2.051)$$

$$P_U(-1.465) = \frac{1}{\sqrt{2\pi}} \int_{-1.758}^{-1.172} e^{-\gamma^2/2} \, d\gamma = 0.0818 = P_U(1.465)$$

$$P_U(-0.879) = \frac{1}{\sqrt{2\pi}} \int_{-1.172}^{-0.586} e^{-\gamma^2/2} \, d\gamma = 0.1556 = P_U(0.879)$$

$$P_U(-0.293) = \frac{1}{\sqrt{2\pi}} \int_{-0.586}^{0} e^{-\gamma^2/2} \, d\gamma = 0.2224 = P_U(0.293) \, .$$

The entropy of this DMS is

$$H(U) = 2.757 \text{ bits/letter}$$

for a distortion of $D = 0.03744$. The rate distortion bound for this distortion is $R(D) = \frac{1}{2} \log(\sigma_s^2/D) = 2.370$ bits/letter. Clearly, the quantizer operates $2.757 - 2.370 = 0.387$ bits/letter above the rate distortion bound.

Comparing the results of the last two examples, we see that the uniform quantizer generates a DMS with a lower entropy (output rate) but a slightly higher distortion. Since the distortion is higher, $R(D)$ is lower and the uniform quantizer operates slightly closer to the rate distortion bound than the MMSE designed quantizer. This demonstrates a well known result that the uniform quantizer may be $R(D)$ optimal under the constraints if entropy coding is used.

5.9   NOTES AND ADDITIONAL REFERENCES

There are several things that are unique about this chapter compared to other chapters on rate distortion theory in the broad coverage information theory and rate distortion theory books. The chapter on rate distortion theory in this book has a strong emphasis on the Shannon lower bound and the conditions for achieving equality in the bound [3, 7]. The Shannon lower bound is one of the most useful tools to a researcher working in lossy source coding and rate distortion theory and deserves this deeper coverage. The inclusion of a section on the Gaussian autoregressive source is also unusual, but since this source model is dominant in speech coding and many other practical source coding applications, it is imperative that there is some understanding of what is known in the rate distortion theory sense about these source models [7, 32]. Further, this chapter has a section on composite source models and conditional rate distortion theory that is not found in other books [7, 33]. Conditional rate distortion theory has been shown to be an invaluable tool for finding the rate distortion bounds for realistic physical sources such as voice and video, and therefore is a must when hoping to have rate distortion theory apply to real sources.

APPENDIX A

Useful Inequalities

In this appendix, we present some common (useful) inequalities that serve as the basis for optimization problem solutions in the several information and rate distortion theory books [4, 5, 14].

We begin with a definition of the important property of convexity. We will use the terminology convex for convex $\cup$ (read cup), and concave for convex $\cap$ (read cap).

Definition A.1 A function $f(x)$ is said to be convex $\cup$ over an interval (a, b) if for every $x_1, x_2 \in (a, b)$ and $0 \le \lambda \le 1$,

$$f(\lambda x_1 + (1 - \lambda)x_2) \le \lambda f(x_1) + (1 - \lambda)f(x_2) \tag{A.1}$$

a function f is said to be strictly convex if equality holds only if $\lambda = 0$ or $\lambda = 1$. A function is

convex over an interval if it always lies below any chord.

Definition A.2 A function f is concave if $-f$ is convex.

Definition A.3 *Jensen's Inequality:* For a random variable X and a convex function f, then

$$Ef(X) \ge f(EX) \,.$$

Jensens's inequality is used in McEliece [14].

Definition A.4 *Information Inequality:*

$$D(p\|q) \ge 0 \,.$$

The Information Inequality is the tool primarily used in Cover and Thomas [5].

Jensen's inequality states that for a random variable W with a distribution defined on an appropriate interval and for a convex function, say $g(x)$, then

$$E[g(W)] \ge g[E(W)]$$

if $E[W]$ exists. If $g(x)$ is concave, the inequality is reversed to yield

$$E[g(W)] \le g[E(W)] \,.$$

Jensen's inequality is particularly useful when working with mutual information, since $I(W; X)$ is a convex function of the transition probabilities for fixed input probabilities, and $I(W; X)$ is a concave function of the input probabilities for fixed transition probabilities.

Gallager [4] often uses the inequality $\log x \leq x - 1$ in his proofs.

APPENDIX B

Laws of Large Numbers

B.1 INEQUALITIES AND LAWS OF LARGE NUMBERS

We present various forms of the laws of large numbers and their related inequalities in this appendix [35, 36].

B.1.1 MARKOV'S INEQUALITY

$$P[X \geq a] \leq \frac{E[X]}{a} \quad \text{for } X \text{ nonnegative}.$$

Proof.

$$
\begin{aligned}
E[X] &= \int_0^a \alpha f_X(\alpha)\, d\alpha + \int_a^\infty \alpha f_X(\alpha)\, d\alpha \\
&\geq \int_a^\infty \alpha f_X(\alpha)\, d\alpha \geq a \int_a^\infty f_X(\alpha)\, d\alpha \\
&= a P[X \geq a],
\end{aligned}
$$

so

$$P[X \geq a] \leq \frac{E[X]}{a}.$$

$\square$

Theorem B.1 (Bienaymé-Chebychev). *Let X be a random variable with $E[X]^r < \infty$ for $r > 0$; then for every $\lambda > 0$, it follows*

$$P[|X| \geq \lambda] \leq \frac{E|X|^r}{\lambda^r}.$$

Proof. The Cumulative Distribution Function for X is $F_X(x)$, so

$$E[|X|^r] = \int_{-\infty}^{\infty} |x|^r dF(x)$$

$$= \int_{|x|<\lambda} |x|^r dF(x) + \int_{|x|\geq\lambda} |x|^r dF(x)$$

$$\geq \int_{|x|\geq\lambda} |x|^r dF(x) \geq \lambda^r \int_{|x|\geq\lambda} dF(x) ,$$

so

$$E[|X|^r] \geq \lambda^r P[|X| \geq \lambda] .$$

$\square$

B.1.2   CHEBYCHEV'S INEQUALITY

For $r = 2$ in the above Theorem,

$$E[|X|^2] \geq \lambda^2 P[|X| \geq \lambda] .$$

Replacing X with $(X - \mu)/\sigma$,

$$P\left[\left|\frac{X - \mu}{\sigma}\right| \geq \lambda\right] \leq \frac{E\left|\frac{X-\mu}{\sigma}\right|^2}{\lambda^2} = \frac{1}{\lambda^2} ,$$

or

$$P[|X - \mu| \geq \lambda\sigma] \leq \frac{1}{\lambda^2} .$$

B.1.3   WEAK LAW OF LARGE NUMBERS

Let $X_1, X_2, \ldots,$ be a sequence of i.i.d. random variables with finite mean $EX = \mu$, then for $\varepsilon > 0 (M_n = \frac{1}{n}\sum_j X_j)$

$$\lim_{n\to\infty} P[|M_n - \mu| < \varepsilon] = 1 .$$

The weak law states that for a large enough fixed values of n, the sample mean using n samples will be close to the true mean with high probability.

The weak law does not address what happens as a function of n. This question is taken up by the strong law.

B.1.4   STRONG LAW OF LARGE NUMBERS

Let $X_1, X_2 \ldots$ be a sequence of i.i.d. random variables with $E[X] = \mu < \infty$ and finite variance, then

$$P\left[\lim_{n \to \infty} M_n = \mu\right] = 1 .$$

This says that w.p. 1 every sequence of sample mean calculations will eventually approach and stay close to $EX = \mu$.

Chebychev's Inequality.

Let Y be a random variable with mean μ and variance σ^2. Then

$$P_r\{|Y - \mu| > \epsilon\} \le \frac{\sigma^2}{\epsilon^2} . \tag{B.1}$$

Weak Law of Large Numbers.

Let $Z_1, Z_2, \ldots, Z_n$, be a sequence of i.i.d. random variables with mean μ and variance σ^2. Let

$$\bar{Z}_n = \frac{1}{n} \sum_{i=1}^{n} Z_i .$$

Then

$$P_r\left\{|\bar{Z}_n - \mu| > \epsilon\right\} \le \frac{\sigma^2}{n\epsilon^2} . \tag{B.2}$$

APPENDIX C

Kuhn-Tucker Conditions

A key result in finding the solution to our several optimization problems is the Kuhn-Tucker conditions, which can be stated as in the following.

Theorem C.1 **[4].** *Let $f(\underline{\alpha})$ be a concave function of $\underline{\alpha} = (\alpha_1, \ldots, \alpha_k)$ over the region $\mathcal{R}$ when $\underline{\alpha}$ is a probability vector. Assume that the partial derivatives, $\partial f(\underline{\alpha})/\partial \alpha_k$ are defined and continuous over the region $\mathcal{R}$ with the possible exception that $\lim_{\alpha_k \to 0} \frac{\partial f(\underline{\lambda})}{\partial \alpha_k}$ may be $+\infty$. Then*

$$\frac{\partial f(\underline{\alpha})}{\partial \alpha_k} = \lambda; \; all \; k \; such \; that \; \alpha_k > 0,$$
$$\frac{\partial f(\underline{\alpha})}{\partial \alpha_k} \leq \lambda; \; all \; k \; such \; that \; \alpha_k = 0,$$

are necessary and sufficient conditions on a probability vector $\underline{\alpha}$ to maximize f over the region $\mathcal{R}$.

Proof. See Gallager [4]. $\qquad\qquad\qquad\qquad\qquad\qquad\qquad\qquad\qquad\qquad\qquad\qquad$ $\square$

Bibliography

[1] C. E. Shannon, "A mathematical theory of communication," *The Bell System Technical Journal*, vol. 27, pp. 379–423, 623–656, 1948. 1, 4, 5, 7, 20, 50, 60, 78

[2] C. E. Shannon and W. Weaver, *The Mathematical Theory of Communication*, 1949. 1, 7

[3] C. E. Shannon, "Coding theorems for a discrete source with a fidelity criterion," *IRE Nat. Conv. Rec*, vol. 4, no. 142-163, 1959. 1, 6, 7, 88, 102

[4] R. G. Gallager, *Information Theory and Reliable Communication*. New York, NY: John Wiley & Sons, Inc., 1968. x, 4, 5, 7, 9, 16, 17, 20, 29, 31, 33, 43, 46, 48, 49, 50, 52, 53, 55, 60, 64, 65, 73, 76, 78, 80, 103, 104, 109

[5] T. M. Cover and J. A. Thomas, *Elements of Information Theory*. New York: Wiley-Interscience, Aug. 1991. x, 4, 5, 6, 7, 9, 16, 17, 25, 27, 29, 31, 33, 37, 43, 46, 48, 49, 50, 52, 53, 55, 60, 62, 63, 64, 67, 70, 73, 76, 78, 80, 81, 85, 88, 96, 103

[6] R. E. Blahut, *Principles and Practice of Information Theory*. Addison-Wesley Longman Publishing Co., Inc., 1987. 4, 6, 7, 25, 48, 64

[7] T. Berger, *Rate Distortion Theory: A Mathematical Basis for Data Compression*. Prentice-Hall, 1971. 6, 7, 9, 17, 81, 83, 84, 85, 88, 89, 93, 95, 96, 102

[8] T. Berger and J. D. Gibson, "Lossy Source Coding," *IEEE Trans. on Information Theory*, vol. 44, no. 6, pp. 2693–2723, Oct. 1998. 6, 8

[9] S. Verdu, V. Anantharam, G. Caire, M. Costa, G. Kramer, and R. Yeung, "Panel on new perspectives for information theory," *Information Theory Workshop, Paraty, Brazil*, Oct. 2011. 7

[10] N. Abramson, *Information Theory and Coding*. McGraw-Hill New York, 1963, vol. 61. 7

[11] R. Fano, *Transmission of Information*. MIT Press, 1961. 7

[12] R. B. Ash, *Information Theory*. Wiley-Interscience, 1965. 7

[13] F. Jelinek, *Probabilistic Information Theory: Discrete and Memoryless Models*. McGraw-Hill, 1968. 7

[14] R. McEliece, *The Theory of Information and Coding*.   Cambridge University Press, 2002. 7, 20, 48, 52, 55, 103

[15] R. M. Gray, *Source Coding Theory*.   Kluwer Academic Publishers, 1990, vol. 83. 8

[16] J. M. Wozencraft and I. M. Jacobs, *Principles of Communication Engineering*.   John Wiley & Sons, Inc., 1965, vol. 1. 8, 69

[17] A. J. Viterbi and J. K. Omura, *Principles of Digital Communication and Coding*.   McGraw-Hill, 1979. 8

[18] K. Sayood, *An Introduction to Data compression*.   Morgan Kaufmann Publishers Inc., 2006. 8, 36, 37, 40, 41

[19] L. D. Davisson and R. M. Gray, *Data Compression*.   Dowden, Hutchinson & Ross, 1976, vol. 14. 8

[20] D. Slepian, *Key Papers in the Development of Information Theory*.   IEEE press, 1974. 8

[21] S. Verdü and S. W. McLaughlin, *Information theory: 50 years of discovery*.   IEEE Press, 2000. 8

[22] J. D. Gibson, T. Berger, T. Lookabaugh, D. Lindbergh, and R. L. Baker, *Digital Compression for Multimedia: Principles and Standards*.   San Francisco, CA: Morgan Kaufmann Publishers Inc., 1998. 8, 36, 40, 43

[23] A. Wyner, "Fundamental limits in information theory," *Proceedings of the IEEE*, vol. 69, no. 2, pp. 239–251, 1981. [Online]. Available: http://ieeexplore.ieee.org/xpls/abs_all.jsp?arnumber=1456224 30

[24] D. A. Huffman, "A method for the construction of minimum-redundancy codes," *Proceedings of the IRE*, vol. 40, no. 9, pp. 1098–1101, 1952. [Online]. Available: http://ieeexplore.ieee.org/xpls/abs_all.jsp?arnumber=4051119 33

[25] J. J. Rissanen and G. G. Langdon, "Arithmetic coding," *IBM Journal of Research and Development*, vol. 23, no. 2, pp. 149–162, 1979. [Online]. Available: http://ieeexplore.ieee.org/stamp/stamp.jsp?arnumber=5390830 40

[26] T. A. Welch, "A technique for high-performance data compression," *Computer*, vol. 17, no. 6, pp. 8–19, 1984. 40, 41

[27] J. Ziv and A. Lempel, "A universal algorithm for sequential data compression," *Information Theory, IEEE Transactions on*, vol. 23, no. 3, pp. 337–343, 1977. [Online]. Available: http://ieeexplore.ieee.org/xpls/abs_all.jsp?arnumber=1055714 40

[28] J. Ziv and A. Lempel, "Compression of individual sequences via variable-rate coding," *Information Theory, IEEE Transactions on*, vol. 24, no. 5, pp. 530–536, 1978. [Online]. Available: http://ieeexplore.ieee.org/xpls/abs_all.jsp?arnumber=1055934 40

[29] C. E. Shannon, "Communication in the presence of noise," *Proceedings of the IRE*, vol. 37, no. 1, pp. 10–21, 1949. [Online]. Available: http://ieeexplore.ieee.org/xpls/abs_all.jsp?arnumber=1697831 67, 80

[30] R. A. McDonald and P. M. Schultheiss, "Information rates of gaussian signals under criteria constraining the error spectrum," *Proceedings of the IEEE*, vol. 52, pp. 415–416, Apr. 1964.

[31] L. D. Davisson, "Rate-distortion theory and application," *Proceedings of the IEEE*, vol. 60, no. 7, pp. 800 – 808, July 1972. 92

[32] R. M. Gray, "Information rates of autoregressive processes," *IEEE Trans. on Information Theory*, vol. 16, no. 4, pp. 412 – 421, Jul. 1970. 93, 102

[33] R. M. Gray, "A new class of lower bounds to information rates of stationary sources via conditional rate-distortion functions," *IEEE Trans. on Information Theory*, vol. IT-19, no. 4, pp. 480–489, July 1973. 95, 96, 102

[34] J. D. Gibson, *Principles of digital and analog communications*.   Macmillan, 1993. 98

[35] A. Leon-Garcia, *Probability and Random Processes for Electrical Engineering*.   Addison-Wesley Longman Publishing Co., Inc., 1989. 105

[36] J. B. Thomas, *An Introduction to Communication Theory and Systems*.   Springer-Verlag, 1988. 105

[37] T. Cover and A. El Gamal, An information theoretic proof of Hadamard's inequality, *IEEE Trans. Inform. Theory*, IT-29 (1983), pp. 930–931. 77

Author's Biography

JERRY GIBSON

Jerry Gibson is Professor of Electrical and Computer Engineering at the University of California, Santa Barbara. He has been an Associate Editor of the *IEEE Transactions on Communications* and the *IEEE Transactions on Information Theory*. He was President of the IEEE Information Theory Society in 1996, and he has served on the Board of Governors of the IT Society and the Communications Society. He was an IEEE Communications Society Distinguished Lecturer for 2007-2008. He is an IEEE Fellow, and he has received The Fredrick Emmons Terman Award (1990), the 1993 IEEE Signal Processing Society Senior Paper Award, the 2009 IEEE Technical Committee on Wireless Communications Recognition Award, and the 2010 Best Paper Award from the *IEEE Transactions on Multimedia*. He is co-author of the books *Digital Compression for Multimedia* (Morgan-Kaufmann, 1998) and *Introduction to Nonparametric Detection with Applications* (Academic Press, 1975 and IEEE Press, 1995) and author of the textbook, *Principles of Digital and Analog Communications* (Prentice-Hall, second ed., 1993). He is Editor-in-Chief of *The Mobile Communications Handbook* (CRC Press, 3rd ed., 2012), Editor-in-Chief of *The Communications Handbook* (CRC Press, 2nd ed., 2002), and Editor of the book, *Multimedia Communications: Directions and Innovations* (Academic Press, 2000).

Printed in the United States
by Baker & Taylor Publisher Services